Undergraduate Topics in Computer Science

Series Editor

Ian Mackie, University of Sussex, Brighton, France

Advisory Editors

Samson Abramsky ⓘ, Department of Computer Science, University of Oxford, Oxford, UK

Chris Hankin ⓘ, Department of Computing, Imperial College London, London, UK

Mike Hinchey ⓘ, Lero—The Irish Software Research Centre, University of Limerick, Limerick, Ireland

Dexter C. Kozen, Department of Computer Science, Cornell University, Ithaca, USA

Hanne Riis Nielson ⓘ, Department of Applied Mathematics and Computer Science, Technical University of Denmark, Kongens Lyngby, Denmark

Steven S. Skiena, Department of Computer Science, Stony Brook University, Stony Brook, USA

Iain Stewart ⓘ, Department of Computer Science, Durham University, Durham, UK

Joseph Migga Kizza, Engineering and Computer Science, University of Tennessee at Chattanooga, Chattanooga, USA

Roy Crole, School of Computing and Mathematics Sciences, University of Leicester, Leicester, UK

Elizabeth Scott, Department of Computer Science, Royal Holloway University of London, Egham, UK

'Undergraduate Topics in Computer Science' (UTiCS) delivers high-quality instructional content for undergraduates studying in all areas of computing and information science. From core foundational and theoretical material to final-year topics and applications, UTiCS books take a fresh, concise, and modern approach and are ideal for self-study or for a one- or two-semester course. The texts are authored by established experts in their fields, reviewed by an international advisory board, and contain numerous examples and problems, many of which include fully worked solutions.

The UTiCS concept centers on high-quality, ideally and generally quite concise books in softback format. For advanced undergraduate textbooks that are likely to be longer and more expository, Springer continues to offer the highly regarded *Texts in Computer Science* series, to which we refer potential authors.

Quentin Charatan · Aaron Kans

Maths For Computing
A Beginner's Guide

Quentin Charatan
School of Architecture, Computing
and Engineering
University of East London
London, UK

Aaron Kans
School of Architecture, Computing
and Engineering
University of East London
London, UK

ISSN 1863-7310 ISSN 2197-1781 (electronic)
Undergraduate Topics in Computer Science
ISBN 978-3-031-69233-8 ISBN 978-3-031-69234-5 (eBook)
https://doi.org/10.1007/978-3-031-69234-5

© The Editor(s) (if applicable) and The Author(s), under exclusive license to Springer Nature Switzerland AG 2025

This work is subject to copyright. All rights are solely and exclusively licensed by the Publisher, whether the whole or part of the material is concerned, specifically the rights of translation, reprinting, reuse of illustrations, recitation, broadcasting, reproduction on microfilms or in any other physical way, and transmission or information storage and retrieval, electronic adaptation, computer software, or by similar or dissimilar methodology now known or hereafter developed.
The use of general descriptive names, registered names, trademarks, service marks, etc. in this publication does not imply, even in the absence of a specific statement, that such names are exempt from the relevant protective laws and regulations and therefore free for general use.
The publisher, the authors and the editors are safe to assume that the advice and information in this book are believed to be true and accurate at the date of publication. Neither the publisher nor the authors or the editors give a warranty, expressed or implied, with respect to the material contained herein or for any errors or omissions that may have been made. The publisher remains neutral with regard to jurisdictional claims in published maps and institutional affiliations.

This Springer imprint is published by the registered company Springer Nature Switzerland AG
The registered company address is: Gewerbestrasse 11, 6330 Cham, Switzerland

If disposing of this product, please recycle the paper.

To Cheeto and Pabla
—Quentin Charatan

To my sister Madhu
—Aaron Kans

Preface

This book is written for students embarking on an undergraduate or foundation degree course in computer science (or related discipline) and aims to provide the basic skills and knowledge of discrete mathematics required for such a course. Whereas many textbooks tend to teach this subject in a way that is more suitable for mathematicians, this text specifically targets first-year students on computing courses and aims to teach only the basic material that they will need for their computing degree.

It is geared towards students who are less confident in mathematics, and topics are taught using everyday language, avoiding lengthy mathematical explanations.

Familiar scenarios are used throughout the text to introduce mathematical concepts, and formal definitions are provided only when a particular concept has been fully explored and understood. Additionally, each chapter discusses the reasons why the particular topic is relevant to the world of computing.

The text covers all the mathematical concepts that are traditionally thought to be necessary for a computing degree but restricts itself only to the material needed for the fundamentals of such a computing programme and does not delve into the higher mathematical concepts; nor does it attempt to cover the more complex areas required for specialisms such as AI or graphical imaging, because it is the authors' experience that these concepts are usually taught on specialist modules when necessary.

Thus, the text begins with set theory, number theory and groups and proceeds to explore relations and functions. It then deals with mathematical logic and proofs, followed by combinatorics and probability. Finally, it covers the essential aspects of graph theory.

Each chapter contains worked examples, as well as end-of-chapter exercises with solutions, which could map ideally onto a lecture/tutorial delivery model.

We would like to thank our publisher, Springer, for the encouragement and guidance that we have received throughout the production of this book. Additionally, we would especially like to thank the computing students at the University of East London for their thoughtful comments and feedback. For support and inspiration, special thanks are due once again to our families and friends.

London, UK
Quentin Charatan
Aaron Kans

Contents

1	**Set Theory**		1
	1.1	Introduction	1
	1.2	What is a Set?	2
	1.3	Notation	2
	1.4	Ordering in a Set	3
	1.5	Repetition in a Set	3
	1.6	Number Sets	3
	1.7	Specifying a Set	3
	1.8	The Universal Set	4
	1.9	The Empty Set	5
	1.10	Cardinality	5
	1.11	Finite Sets and Infinite Sets	5
	1.12	Subsets	6
	1.13	Proper Subsets	6
	1.14	Set Operations	7
		1.14.1 Union	8
		1.14.2 Intersection	8
		1.14.3 Difference	9
		1.14.4 Complement	9
		1.14.5 Cartesian Product	10
	1.15	Venn Diagrams	11
	1.16	Symmetric Difference	13
	1.17	De Morgan's Law	16
	1.18	Classes of Sets and Power Sets	16
	1.19	Russell's Paradox	17
	1.20	Application to Computing	18
	1.21	Exercises	18
2	**Sets and Groups**		21
	2.1	Introduction	21
	2.2	The Algebra of Sets	22
	2.3	Number Types	23
		2.3.1 Natural Numbers	23

		2.3.2	Integers	23
		2.3.3	Rational Numbers	23
		2.3.4	Real Numbers	24
		2.3.5	Complex Numbers	24
	2.4	Number Types and Programming		24
	2.5	Working with Complex Numbers		25
	2.6	Number Sets		26
	2.7	Discrete Versus Continuous Values		26
	2.8	Countable and Non-countable Sets		26
	2.9	Operations		27
		2.9.1	Commutative Operations	27
		2.9.2	Associative Operations	28
	2.10	Groups		30
	2.11	Notation and Terminology		32
	2.12	Relevance to Computing		35
	2.13	Exercises		35
3	**Relations**			37
	3.1	Introduction		37
	3.2	Relations: Definition and Notation		37
	3.3	Inverse Relations		40
	3.4	Relations on a Set		41
	3.5	Digraphs of Relations		41
	3.6	Symmetric Relations		41
	3.7	Reflexive Relations		42
	3.8	Transitive Relations		43
	3.9	Equivalence Relations		43
	3.10	Equivalence Classes		44
	3.11	Binary Relations		45
	3.12	n-Ary Relations		45
	3.13	Relevance to Computing: Relational Databases		46
	3.14	Exercises		47
4	**Functions**			49
	4.1	Introduction		49
	4.2	Definition		49
	4.3	Functions as Mappings		50
		4.3.1	Practical Example	51
		4.3.2	Using the Password Function	52
	4.4	A Function as an Input/Output Device		52
	4.5	Functions as Formulae		52
		4.5.1	Applying our Function	52

		4.5.2	Another Example	53
	4.6	Functions as Equations		56
	4.7	The Signature of a Function		56
	4.8	Specifying a Function		56
	4.9	Specifying Formulaic Functions		57
	4.10	Not all Mappings are Functions		57
	4.11	Functions with More than One Input		58
	4.12	Function Composition		59
	4.13	Injective Functions		60
	4.14	Surjective Functions		60
	4.15	Bijective Functions		60
	4.16	Application to Computing		61
		4.16.1	Computer Programming	61
		4.16.2	Business Software	62
	4.17	Exercises		62
5	**Propositional Logic**			65
	5.1	Introduction		65
	5.2	Mathematical Logic		66
	5.3	Propositions		66
	5.4	Logical Operators (Connectives)		66
		5.4.1	The AND operator ($\wedge$)	66
		5.4.2	The OR Operator ($\vee$)	67
		5.4.3	The NOT Operator ($\neg$)	68
	5.5	Constructing Truth Tables		70
	5.6	Logical Equivalence		70
	5.7	De Morgan's Law		70
	5.8	Commutativity and Associativity		72
	5.9	The Implication Operator ($\Rightarrow$)		72
	5.10	The Equivalence Operator ($\Leftrightarrow$)		73
	5.11	The Equivalence Operator Versus Logical Equivalence		74
	5.12	Order of Precedence of Logical Operators		75
	5.13	Tautologies and Contradictions		75
	5.14	The EXCLUSIVE OR Operator ($\oplus$)		76
	5.15	Converse, Inverse and Contrapositive		76
	5.16	Algebra of Propositions		77
	5.17	Some More Algebra		78
	5.18	Application to Computing		79
		5.18.1	Programming	79
		5.18.2	Digital Electronics and Logic Gates	80
	5.19	Three-Valued Logic		81
	5.20	Notation in other Texts		82
	5.21	Exercises		82

6	**Predicate Logic and Proofs**		85
	6.1	Introduction	85
	6.2	Predicate Logic	86
	6.3	Examples of Predicates	86
	6.4	The Domain of Discourse	86
	6.5	Giving Values to the Variables	87
		6.5.1 Substitution	87
		6.5.2 Quantification	87
		6.5.3 Negating Quantified Predicates	90
	6.6	Proof by Natural Deduction	91
		6.6.1 Modus Ponens	91
		6.6.2 Modus Tollens	91
		6.6.3 The Chain Rule	92
		6.6.4 Alternative Names	93
		6.6.5 Some Other Laws	93
	6.7	Application to Computing	96
	6.8	Proof by Induction	96
	6.9	Exercises	99
7	**Matrices**		101
	7.1	Introduction	101
	7.2	Definition and Examples	101
	7.3	Matrix Operations	102
		7.3.1 Transposition	102
		7.3.2 Addition and Subtraction of Matrices	102
		7.3.3 Scalar Multiplication	103
		7.3.4 Matrix Multiplication	103
	7.4	The Determinant of a Matrix	105
		7.4.1 Calculating the Determinant of a 3×3 Matrix	106
	7.5	Identity Matrices	107
	7.6	The Inverse of a Matrix	107
		7.6.1 Finding the Inverse	108
	7.7	Application to Computing	110
	7.8	Using Matrices to Solve Linear Equations	110
	7.9	Row Operations on a Matrix	112
	7.10	Solving Equations Using the Gauss-Jordan Elimination Method	113
	7.11	Exercises	114

8	**Combinatorics**		117
	8.1	Introduction	117
	8.2	Placing Items in Order	118
	8.3	Factorials	118
	8.4	Choosing just Some of the Items	119
	8.5	The Formula for Permutations	120
	8.6	Combinations	120
	8.7	The Formula for Combinations	120
	8.8	What Is the Value of 0!?	121
	8.9	Allowing Repetition (When Order Is Important)	121
	8.10	Allowing Repetition (When Order Is Not Important)	121
	8.11	Summary of Formulae	122
	8.12	Deriving the Formula for Choosing Items when Order Is Not Important and Repetition Is Allowed	122
	8.13	Pascal's Triangle	126
	8.14	Binomial Expansion	127
	8.15	Application to Computing	129
	8.16	Exercises	129
9	**Probability**		131
	9.1	Introduction	131
	9.2	Terminology and Definitions	132
	9.3	Outcomes, Sample Space and Events	132
	9.4	Calculating Probability	133
	9.5	Probability that One or Another Event Happens	135
	9.6	Mutually Exclusive Events	135
		9.6.1 The Addition Rule	136
		9.6.2 The Addition Rule for Mutually Exclusive Events	137
	9.7	Probability Distribution	138
		9.7.1 Non-uniform Probability Distribution	139
	9.8	Independent Events	141
	9.9	Random Variables	142
	9.10	Expected Value (Mean Value)	143
	9.11	Conditional Probability	145
		9.11.1 Tree Diagrams	145
	9.12	Bayes' Theorem	147
	9.13	Binomial Probability	150
		9.13.1 The Binomial Probability Formula	151
	9.14	Application to Computing	153
	9.15	Exercises	153

10	**Graph Theory**		157
	10.1	Introduction	158
	10.2	Definitions	158
	10.3	Multigraphs	159
	10.4	Connected and Unconnected Graphs	159
	10.5	Degree of a Vertex	160
		10.5.1 Sum of Degrees of Vertices Theorem	160
	10.6	Distance Between Two Vertices	160
	10.7	Eccentricity of a Vertex	161
	10.8	Radius and Diameter	162
	10.9	Subgraphs	163
	10.10	Paths, Trails, Circuits and Cycles	163
	10.11	Isomorphic and Homeomorphic Graphs	165
	10.12	Traversable Graphs	166
		10.12.1 Eulerian Graphs	167
		10.12.2 Hamiltonian Graphs	167
	10.13	Weighted Graphs	169
	10.14	Trees	170
		10.14.1 Spanning Trees	171
		10.14.2 Binary Trees	173
	10.15	Planar Graphs	178
	10.16	Directed Graphs	180
	10.17	Application to Computing	183
	10.18	Exercises	184
11	**Solutions to Exercises**		189
	11.1	Chapter 1	189
	11.2	Chapter 2	190
	11.3	Chapter 3	193
	11.4	Chapter 4	194
	11.5	Chapter 5	194
	11.6	Chapter 6	196
	11.7	Chapter 7	198
	11.8	Chapter 8	200
	11.9	Chapter 9	204
	11.10	Chapter 10	210
Index			213

Set Theory

At the end of this chapter you should be able to:

- give a definition of a set and provide examples of sets;
- recognise and utilise the notation that forms the body of set theory;
- specify a set using set comprehension;
- explain the meaning of the *universal set* and the *empty set*;
- determine the cardinality of a set;
- explain the meaning of the terms *subset* and *proper subset*;
- perform standard set operations and recognise the various set operators;
- use Venn diagrams to represent sets;
- provide a statement of De Morgan's law;
- find the power set of a given set.

1.1 Introduction

In the world of computing we very often have to write applications that handle *collections*. These could be collections of customers, bank accounts, students, patients, network users, invoices and many many more.

In this chapter we will be studying set theory, and as we shall see, a set is a particular type of collection; we will be exploring ways of defining sets and performing operations on sets, and in the next chapter we will go on to study a number of laws associated with set theory.

As we progress, you will see that an understanding of set theory is extremely important when studying other areas of mathematics such as probability, which you will encounter in this text.

1.2 What is a Set?

A **set** is an *unordered* collection of distinct objects. These objects are called the **elements** or **members** of the set. The objects can be anything—people, words, numbers, animals or even other sets.

Examples of possible sets could be:

- students at a particular university;
- positive integers less than ten;
- countries in the European Union;
- words in the English language.

1.3 Notation

We often use upper-case letters to refer to sets. When we list the members of a set we enclose them in curly brackets, separating each element of the set with a comma.

Examples
If A is the set of all the countries in the UK, we could write:

$$A = \{\text{WALES, SCOTLAND, ENGLAND, NORTHERN IRELAND}\}$$

If B is the set of even numbers greater than zero and less than ten, then:

$$B = \{2, 4, 6, 8\}$$

When reasoning about sets in mathematics, we often use lower case letters to represent the elements of a set. For example, we might have:

$$C = \{a, d, f, e, x, g\}$$

The following notation is used to indicate that an object is a member of a set:

$x \in A$ means x is an element of A.
$x \notin A$ means x is not an element of A.

Two sets are said to be equal if and only if they contain exactly the same elements (no more and no fewer).

1.4 Ordering in a Set

A set is an *unordered* collection of elements. Therefore the order in which the elements is listed in a set is irrelevant. There is no notion of "first element" or "second element" etc.

Example

$$\{a, b, c, d, e\} = \{e, d, b, c, a\}$$

These two sets are identical—they contain exactly the same elements. We can list them in any order we choose.

1.5 Repetition in a Set

Elements of a set are named only once.

If A is the set $\{1, 2, 3\}$, we could add 3 to this set as many times as we wanted, and the set would remain unchanged.

1.6 Number Sets

It is common to refer to the sets of numbers by the following letters:

$\mathbb{N}$ is the set of natural numbers (positive whole numbers, including zero)
$\mathbb{Z}$ is the set of integers (positive and negative whole numbers, including zero)
$\mathbb{Q}$ is the set of rational numbers (this will be explained in the next chapter)
$\mathbb{R}$ is the set of real numbers (decimal numbers)
$\mathbb{C}$ is the set of complex numbers (this will be explained in the next chapter).

1.7 Specifying a Set

There are three ways to specify a set:

1. By listing the elements

$$A = \{\text{CAT, DOG, HORSE, MOUSE}\}$$
$$B = \{a, c, g, m, z\}$$

2. By describing the elements

$$S = \{\textit{Students at The University of East London}\}$$

$$E = \{\textit{even numbers greater than } 0 \textit{ and less than } 10\}$$

3. By comprehension

$$A = \{x \in \mathbb{N} | x < 10\}$$

The bar (|) is read as "such that". So this reads:
 The set of x's which are elements of $\mathbb{N}$ (natural numbers) such that x is less than 10.
 So in this case: $A = \{0, 1, 2, 3, 4, 5, 6, 7, 8, 9\}$.

Worked Example 1
Express the following specification of a set M in words:

$$M = \{x \in \mathbb{R} | x \geq 100\}$$

Solution
M is the set of real numbers greater than or equal to 100.

Worked Example 2
Using set comprehension, specify a set D that contains all the negative integers.

Solution

$$D = \{x \in \mathbb{Z} | x < 0\}$$

1.8 The Universal Set

Consider the following set:

{ROSE, LILY, IRIS, HYACINTH, VIOLET}

What kind of objects are these?
Flowers?
Names?
Words in the English language?
 It is important to know what sort of objects a set contains. When we specify a set, we should always be aware of a bigger set from which our set takes its objects. This bigger set is called the **universal set**. It is usually given the letter U.

Examples

- If we take the set of students on a particular course, the universal set could be the set of people.
- If we take the set {ELEPHANT, LION, DOG}, the universal set could be the set of animals.
- If we have a set of positive whole numbers, the universal set could be natural numbers. But it could also be integers, or real numbers. We need to make it clear which one it is.

1.9 The Empty Set

It is possible for a set to contain no members. A set that contains no members is called **the empty set** and is referred to by this symbol: ∅

Note
A set containing only one member is often referred to as a **unitary** set.

1.10 Cardinality

The number of elements that a set contains is called the **cardinality** of that set.
There are two ways of referring to the cardinality of a set A.

Either: $n(A)$.

Or: $|A|$
In this text we will use the first one.

Examples

$$A = \{a, b, c, d, e, f\} \quad n(A) = 6$$
$$B = \{1, 4, 6\} \quad n(B) = 3$$
$$C = \emptyset \quad n(C) = 0$$
$$D = \{0\} \quad n(D) = 1$$

1.11 Finite Sets and Infinite Sets

In the previous examples we were able to state the cardinality of the sets easily. This is because when we counted the members, the counting came to an end and we had a final result.

Such sets are called **finite** sets.

However there are certain sets where this is not so straightforward, because we never finish counting. Such sets are called **infinite** sets.

All of the number sets that we referred to earlier are infinite sets.

Worked Example 3
State whether each of the following sets is finite or infinite:

(a) The set of positive odd integers greater than 10.
(b) The set of positive odd integers less than 10.
(c) The set of all the people in the world.

Solution

(a) Infinite—it will range from 11 up to infinity.
(b) Finite—it is the set $\{1, 3, 5, 7, 9\}$
(c) Finite.

1.12 Subsets

Consider a set A and a larger set B.

If all the elements contained in A are also elements of B, then A is a **subset** of B.

The following notation is used to mean A is a subset of B: $A \subset B$.
The following means A is *not* a subset of B: $A \not\subseteq B$.

Example
Consider the following sets:

$$C = \{\text{RED, BLUE, YELLOW, GREEN, PURPLE}\}$$
$$A = \{\text{RED, BLUE, GREEN}\}$$
$$B = \{\text{RED, BLUE, YELLOW, PINK}\}$$

In this case $A \subset C$ but $B \not\subseteq C$.

1.13 Proper Subsets

Strictly speaking the expression:

$$A \subset B$$

means A is a *proper* subset of B. This means that we exclude the possibility that A could be equal to B. In other words, there is always at least one more element in B than there is in A. If we want to include the possibility that the two sets are equal, then we use the symbol $\subseteq$.

So: $A \subseteq B$ means that A is a subset of B or is equal to B.

Worked Example 4

Consider the following sets of natural numbers:

$$A = \{1, 4, 6, 7, 9, 10\}$$
$$B = \{6, 7, 9, 10\}$$
$$C = \{3, 7, 9, 10\}$$
$$D = \{10, 7, 6, 9\}$$

For each of the following, state whether the expression is true or false:

(a) $B \subset A$
(b) $A \subset B$
(c) $B = D$
(d) $C \not\subset A$
(e) $D \subset B$
(f) $B \subseteq A$

Solution

(a) True
(b) False
(c) True
(d) True
(e) False
(f) True

1.14 Set Operations

In arithmetic, the basic operations we can perform are:

- addition
- subtraction
- multiplication
- division.

We are now going to study the following operations that we can perform on sets:

- union
- intersection
- difference
- complement
- Cartesian product.

1.14.1 Union

The **union** of two sets, A and B is the set which contains all the elements of A and all the elements of B. We use the following notation:

$$A \cup B \text{ means } A \text{ union } B$$

Example

if $A = \{\text{JOHN, DELROY, ADEWALE, MOHAMMED}\}$.
and $B = \{\text{JOHN, SHEILA, DELROY, ZELDA}\}$.
then $A \cup B = \{\text{JOHN, SHEILA, ADEWALE, MOHAMMED, DELROY, ZELDA}\}$.

Note
Although John and Delroy appear in both sets, they only appear once in the final set: as we know, we do not record duplicates in a set.

1.14.2 Intersection

The **intersection** of two sets, A and B is the set which contains all the elements that are common to both A and B.
We use the following notation:

$A \cap B$ means A intersection B

Example

if $A = \{\text{JOHN, DELROY, ADEWALE, MOHAMMED}\}$.
and $B = \{\text{JOHN, SHEILA, DELROY, ZELDA}\}$.
then $A \cap B = \{\text{JOHN, DELROY}\}$.

Note
If two sets have no common elements, then the intersection is the empty set.
So:

if $X = \{a, b, d, e, g\}$.
and $Y = \{m, n\}$.
then $X \cap Y = \emptyset$.

1.14.2.1 The Exclusion Principle
When we find the union of two sets, we don't include duplicates.

1.14 Set Operations

So the cardinality of the union is found be adding the cardinality of the two sets and subtracting the cardinality of the intersection (otherwise we would be counting the common elements twice).

$$n(A \cup B) = n(A) + n(B) - n(A \cap B)$$

For example:

if $A = \{a, b, d, e, g\}$ $n(A) = 5$
and $B = \{a, b, c, f\}$ $n(B) = 4$
$A \cap B = \{a, b\}$ $n(A \cap B) = 2$
$n(A \cup B) = 5 + 4 - 2 = 7$

This is correct because: $A \cup B = \{a, b, d, e, g, c, f\}$ so $n(A \cup B) = 7$.

1.14.3 Difference

The **difference** of two sets, A and B, is the set which contains the elements that belong to A but do not belong to B.

We use the following notation:

$A \backslash B$ means A difference B.

Example

if $A = \{$JOHN, DELROY, ADEWALE, MOHAMMED$\}$.
and $B = \{$JOHN, SHEILA, DELROY, ZELDA$\}$.
then $A \backslash B = \{$ADEWALE, MOHAMMED$\}$.

1.14.4 Complement

The **complement** of a set A is the set which contains all the elements in the universal set except for those that belong to A.

There are several different notations for the complement of A:

$\overline{A}$ or A' or A^c

We will use the first one.

Example
If the universal set, U, is all the students at the University of East London.

and $A = \{$All students who study computing at the University of East London$\}$.
then $\overline{A} = \{$All the students at the University of East London who do not study computing$\}$.

Another way of describing the complement is: $\overline{A} = U \setminus A$
You should be able to see that: $A \cup \overline{A} = U \quad A \cap \overline{A} = \emptyset$
and: $\overline{\emptyset} = U \quad \overline{U} = \emptyset$.

Note
In some texts you will see this described as the *absolute complement*, while the difference is referred to as the *relative complement*.

1.14.5 Cartesian Product

The Cartesian product of a set A and a set B is written as: $A \times B$.
It is best explained by an example:

if $A = \{a, b, c\}$ and $B = \{d, e\}$.
Then $A \times B = \{(a, d), (a, e), (b, d), (b, e), (c, d), (c, e)\}$.

The Cartesian product results in a set of all possible **ordered pairs**, made up of one element from the first set, followed be one element from the second.

Worked Example 5
Consider the following sets:

$$A = \{b, d, f, g, h, x, y\}$$
$$B = \{f, g\}$$
$$C = \{g, x, z\}$$
$$D = \{z\}$$

(a) Evaluate the following:
   (i) $A \cup C$
   (ii) $B \cup C$
   (iii) $B \setminus C$
   (iv) $B \cap D$
   (v) $B \times C$
(b) If the universal set is $\{b, d, f, g, h, x, y, z, w\}$, what is the value of $\overline{A}$?

Solution

(a)
    (i) $A \cap C = \{g, x\}$
    (ii) $B \cup C = \{f, g, x, z\}$
    (iii) $B \setminus C = \{f\}$
    (iv) $B \cap D = \emptyset$
    (v) $B \times C = \{(f, g), (f, x), (f, z), (g, g), (g, x), (g, z)\}$
(b) $\overline{A} = \{z, w\}$.

1.15 Venn Diagrams

A Venn diagram is a way to represent sets pictorially. The universal set is represented by a rectangle, and a set is represented by a circle. Examples are shown in Figs. 1.1, 1.2, 1.3, 1.4, 1.5 and 1.6.

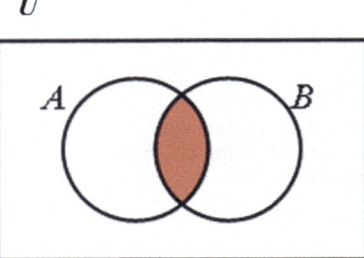

Fig. 1.1 The shaded area represents the intersection of A and B, $A \cap B$

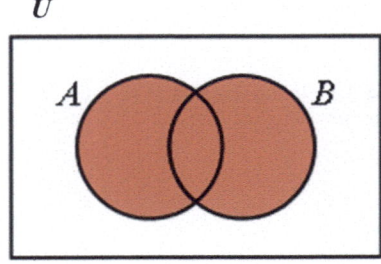

Fig. 1.2 The shaded area represents the union of A and B, $A \cup B$

Fig. 1.3 The shaded area represents A difference B, A\B

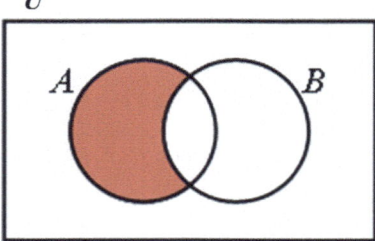

Fig. 1.4 The shaded area represents the complement of A, $\overline{A}$

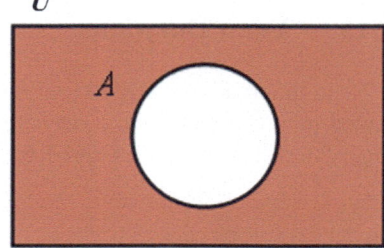

Fig. 1.5 In this diagram, A and B, have no common elements. They are said to be **disjoint**. $A \cap B = \emptyset$

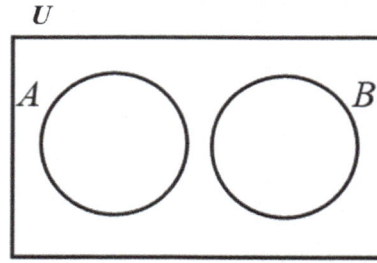

Fig. 1.6 In this diagram, A is a proper subset of B. $A \subset B$

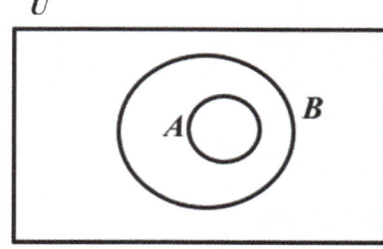

1.16 Symmetric Difference

Symmetric difference, represented by the symbol Δ (or sometimes $\oplus$) is defined as:

$$A \Delta B = (A \backslash B) \cup (B \backslash A)$$

The shaded area in the Venn diagram in Fig. 1.7 represents the symmetric difference:

Worked Example 6
Consider the following sets:

$$A = \{a, b, c, d, e, f\}$$
$$B = \{x, b, c, d, y, w, z\}$$

The universal set, $U = \{a, b, c, d, e, f, x, y, w, z, p, q, r\}$.
Represent this information on a Venn diagram.

Solution
See Fig. 1.8.

Fig. 1.7 Symmetric difference

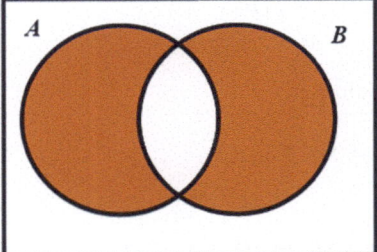

Fig. 1.8 Solution to Worked example 6

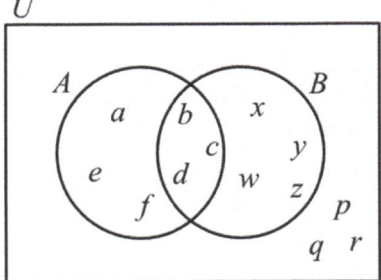

Fig. 1.9 Solution to Worked example 7(a)

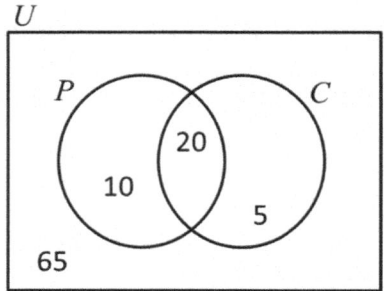

Worked Example 7
This question refers to students at a particular sixth-form college, where there is total of 100 students.

P is the set of students who take physics, and C is the set of students who take chemistry.

30 students take physics, and 25 take chemistry. 20 students take both.

(a) Represent this information on a Venn diagram.
(b) Give values for the following:
   (i) $n(P \cap C)$
   (ii) $n(P \cup C)$
   (iii) $n(P \backslash C)$
   (iv) $n(\overline{P \cup C})$

Solution

(a) See Fig. 1.9
(b)
   (i) $n(P \cap C) = 20$
   (ii) $n(P \cup C) = 35$
   (iii) $n(P \backslash C) = 10$
   (iv) $n(\overline{P \cup C}) = 65$.

Worked Example 8
This question refers the same college as the previous question, where there are 100 students.

P is the set of students who take physics, and C is the set of students who take chemistry, and M is the set of students who take mathematics.

30 students take physics, and 25 take chemistry and 22 take mathematics.

20 students take both physics and chemistry. 18 students take physics and mathematics. 15 students take chemistry and mathematics.

1.16 Symmetric Difference

Fig. 1.10 Solution to Worked example 8(a)

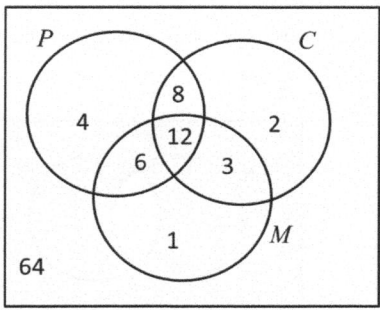

12 students take physics, chemistry and mathematics.

(a) Represent this information on a Venn diagram.
(b) Give values for the following:
   (i) $n(P \cap C \cap M)$
   (ii) $n(P \cup C \cup M)$
   (iii) $n(P \backslash M)$
   (iv) $n(\overline{P \cap C \cap M})$

Solution

(a) See Fig. 1.10.
(b)
   (i) $n(P \cap C \cap M) = 12$
   (ii) $n(P \cup C \cup M) = 36$
   (iii) $n(P \backslash M) = 12$
   (iv) $n(\overline{P \cap C \cap M}) = 88$

Worked Example 9
Use Venn diagrams to prove the following:

$$\overline{A \cap B} = \overline{A} \cup \overline{B}$$

See Fig. 1.11.

This is the left hand side of the equation:

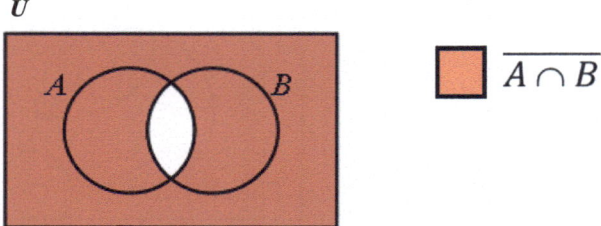

This is the right hand side of the equation:

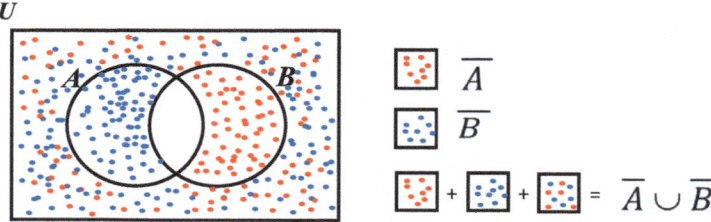

We can see that in both diagrams the shaded areas are the same.

Fig. 1.11 Solution to worked example 9

1.17 De Morgan's Law

In the last slide we used Venn Diagrams to show that:

$$\overline{A \cap B} = \overline{A} \cup \overline{B}$$

It can also be shown that:

$$\overline{A \cup B} = \overline{A} \cap \overline{B}$$

These two identities are two versions of an important law in set theory known as **De Morgan's Law.**

1.18 Classes of Sets and Power Sets

In set theory we often use the word *class* to describe a collection of sets (or a set of sets). For example the class C might be defined as follows:

$$C = \{\{1, 2, 3\}, \{4, 5\}, \{1, 4, 6\}\}$$

The **power set** of a set A is the class of all the subsets of A (including the empty set and A itself). This is written as:

$$P(A)$$

Example

if: $A = \{CAT, DOG, PARROT\}$.

Then:

$$P(A) = \{\emptyset, \{CAT\}, \{DOG\}, \{PARROT\}, \{CAT, DOG\}, \{CAT, PARROT\},\\ \{PARROT, DOG\}, \{CAT, DOG, PARROT\}\}$$

Note

If a set has a cardinality of n, then the number of elements in the power set is 2^n.

Worked Example 11

If A is the set $\{x, y\}$, what is the power set, $P(A)$?

Solution

$$P(A) = \{\emptyset, \{x\}, \{y\}, \{x, y\}\}$$

Worked Example 12

(a) If a set has a cardinality of 5, then how many elements will be in the power set?
(b) How many **proper** subsets does the above set have?

Solution

(a) Number of subsets $= 2^5 = 32$
(b) The number of proper subsets will be one fewer, 31 (because it does not include the set itself).

1.19 Russell's Paradox

The British philosopher Bertrand Russell realised that there was a potential problem with the notion of a "set of sets". This is described below:

If T is the set of all the sets which are not elements of themselves, then is T itself an element of T?

If T is an element of T then our definition of T no longer stands, therefore T cannot be an element of T.

However, if T is *not* an element of T, then by definition it must be an element of T!

This is known as Russell's Paradox. Using the word *class* in this context instead of *set*, is a neat way of avoiding Russell's Paradox.

1.20 Application to Computing

Different types of collections such as sequences and sets occur very frequently in every day situations which, as computer scientists, we need to model in our applications. However, the importance of set theory in computing lies not only in direct application, but also in other areas of mathematics that are central to the understanding of computer science. We will see in future chapters how set theory is vital to our understanding of functions, of logic and of probability all of which are of huge importance to computer science.

1.21 Exercises

1. Express the following specification of a set M in words:

$$M = \{x \in \mathbb{N} | x \geq 50\}$$

2. Using set comprehension, specify a set A that contains all the integers greater than -5 and less than 5.
3. State whether each of the following sets is finite or infinite:
   (a) The set of natural numbers between 50 and 100.
   (b) The set of real numbers less than 10.
   (c) The set of chemical elements discovered so far.
4. Consider the following sets:

$$A = \{a, b, d, e, g, h, x\}$$
$$B = \{a, b, c, d\}$$
$$C = \{g, h, x, a\}$$
$$D = \{h, x, a, g\}$$

   For each of the following, state whether the expression is true or false:
   (a) $C \subset A$ (b) $A \subset C$ (c) $C = D$ (d) $B \not\subset A$ (e) $D \subset C$
   (f) $C \subseteq D$
5. Consider the following sets:

$$A = \{\text{APPLE, ORANGE, PEAR, BANANA, PLUM, LEMON}\}$$

1.21 Exercises

$B = \{\text{APPLE, MANGO, ORANGE}\}$
$C = \{\text{ORANGE, GRAPE, CHERRY}\}$
$D = \{\text{BANANA}\}$

(a) Evaluate the following:
    (i) $A \cup B$ (ii) $B \cap C$ (iii) $A \backslash B$ (iv) $B \cap D$ (v) $B \times D$ (vi) $n(C)$

(b) If the universal set is:

{APPLE, ORANGE, PEAR, BANANA, PLUM, LEMON, MANGO, GRAPE, CHERRY, PINEAPPLE},

what is the value of $\overline{A}$?

6. $A = \{a, b, c, d, e\}$ $B = \{f, d, e, g, h\}$
The universal set $U = \{a, b, c, d, e, f, g, h, i, j\}$.
Represent this information on a Venn diagram.

7. Consider the following sets:

$A = \{\text{APPLE, ORANGE, PEAR, BANANA, PLUM, LEMON}\}$
$B = \{\text{APPLE, MANGO, ORANGE}\}$

Evaluate the following:
$A \triangle B$

8. This question refers to 30 people who were surveyed about the type of vehicles they own.
B is the set of people who own bicycles, and C is the set of people who own cars.
15 people own bicycles, and 12 own cars. 4 people own both.
(a) Represent this information on a Venn diagram.
(b) Give values for the following:
    (i) $n(B \cap C)$ (ii) $n(B \cup C)$ (iii) $n(B \backslash C)$ (iv) $n(\overline{B \cup C})$

9. By drawing a Venn diagram show that: $A \backslash B = A \cap \bar{B}$

10. If A is the set $\{x, y, z\}$, what is the power set, $P(A)$?

11.
(a) If a set has a cardinality of 4, then how many elements will be in the power set?
(b) How many **proper** subsets does the above set have?

Sets and Groups

At the end of this chapter you should be able to:

- solve problems using set algebra;
- distinguish between **natural numbers**, **integers**, **rational numbers**, **real numbers** and **complex numbers**;
- distinguish between **discrete** and **continuous** values;
- determine whether an operation is **commutative** and/or **associative**;
- perform simple algebra using complex numbers;
- define the meaning of the term **group** and state the criteria that determine whether a set is a group under a particular operation;
- use the above criteria to decide whether a set and an operation constitute a group;
- explain the meaning of the term **Abelian group**;
- explain the terms **semigroup** and **monoid**.

2.1 Introduction

In this chapter we expand on the work we did in chapter 1, and describe some laws of set algebra which we can use to solve problems. We apply our knowledge of sets to sets of numbers, and learn to distinguish between different types of numbers. Finally we explore the important mathematical concept of a **group**.

2.2 The Algebra of Sets

There are a number of laws that follow directly from the definitions that we covered in the last chapter:

Idempotent Laws	Identity Laws	Complement Laws
$A \cup A = A$	$A \cup \emptyset = A$	$A \cup \overline{A} = U$
$A \cap A = A$	$A \cup U = U$	$A \cap \overline{A} = \emptyset$
	$A \cap \emptyset = \emptyset$	$\overline{\emptyset} = U$
	$A \cap U = A$	$\overline{U} = \emptyset$

Two other laws which are not quite so obvious are:

Distributive law	De Morgan's law
$A \cup (B \cap C) = (A \cup B) \cap (A \cup C)$	$\overline{A \cup B} = \overline{A} \cap \overline{B}$
$A \cap (B \cup C) = (A \cap B) \cup (A \cap C)$	$\overline{A \cap B} = \overline{A} \cup \overline{B}$

It can be seen that the above pairs are duals of each other. This is a fact of set algebra called **the principle of duality**.

Worked Example 1
Show that:
$$A \cap (\overline{A} \cup B) = A \cap B$$

Solution

$$\begin{aligned}
& A \cap (\overline{A} \cup B) \\
&= (A \cap \overline{A}) \cup (A \cap B) \quad \text{Distributive Law} \\
&= \emptyset \cup (A \cap B) \quad \text{Complement Law} \\
&= A \cap B \quad \text{Identity Law}
\end{aligned}$$

Worked Example 2
Show that:
$$(A \cap B) \cup (A \cap \overline{B}) = A$$

Solution

$$\begin{aligned}
& (A \cap B) \cup (A \cap \overline{B}) \\
&= A \cap (B \cup \overline{B}) \quad \text{Distributive Law} \\
&= A \cap U \quad \text{Complement Law} \\
&= A \quad \text{Identity Law}
\end{aligned}$$

2.3 Number Types

Numbers can be categorised into five distinct types as we saw in the previous chapter.

2.3.1 Natural Numbers

Natural numbers are the numbers that we use for counting. That is to say, whole numbers from 0 to infinity.

Examples of natural numbers are:

$$3, 10, 289, 1098.$$

2.3.2 Integers

Integers include the natural numbers, but also include the numbers less than zero (negative numbers). Integers are therefore whole numbers from minus infinity to plus infinity.

Examples of integers are:

$$-1, 34, -235, 0, 195.$$

2.3.3 Rational Numbers

Rational numbers are numbers that can be expressed as:

$$\frac{p}{q}$$

where p and q are integers.

Some numbers occur naturally in nature but cannot be expressed as rational numbers—they are **irrational** numbers (also called **surds**). Examples of irrational numbers are:

- $\sqrt{2}$
- π (the number of times the diameter of a circle goes into the circumference)
- Euler's number, e, which is a very important mathematical constant.

2.3.4 Real Numbers

Real numbers are numbers that have a fractional part. They include rational numbers but also include irrational numbers. They also include integers (and therefore natural numbers) because integers can be expressed as, for example, 2.0, 3.0, − 5.0 and so on.

Examples of real numbers are:

$$-3.127, 34.987, 0.001, -108.7, 3.0.$$

2.3.5 Complex Numbers

Complex numbers are used in advanced mathematics and consist of a *real* part and an *imaginary* part. The imaginary part contains a special number i, which is defined as:

$$\sqrt{-1}$$

Of course there is no real number that gives − 1 when squared, which is why we use the term "imaginary number".

A typical complex number might look like this: $3 + 2i$.

The set of complex numbers also contains real numbers (the coefficient of i is 0).

Worked Example 3
Consider the following numbers:

−4.82   33   − 262   45.987   0.3

Which of these numbers are:
(a) Real numbers   (b) Integers   (c) Natural numbers

Solution

(a) All of them are real numbers.
(b) 33 and − 262 are integers (as well as being real numbers)
(c) 33 is a natural number (as well as being an integer and a real number).

2.4 Number Types and Programming

In most programming languages we have to declare the type of a variable before we use it. This is because different number types are stored in different ways in memory. It takes up a lot more memory to store a real number than an integer for instance. As an example, the available types for the Java programming language are shown in Table 2.1.

2.5 Working with Complex Numbers

Table 2.1 Java number types

Java type	Range of values	Allows for
Byte	Very small integers	-128 to 127
Short	Small integers	$-32{,}768$ to $32{,}767$
Int	Big integers	$-2{,}147{,}483{,}648$ to $2{,}147{,}483{,}647$
Long	Very big integers	$-9{,}223{,}372{,}036{,}854{,}775{,}808$ to $9{,}223{,}372{,}036{,}854{,}775{,}807$
Float	Real numbers	$\pm 1.4 * 10^{-45}$ to $3.4 * 10^{38}$
Double	Very big real numbers	$\pm 4.9 * 10^{-324}$ to $1.8 * 10^{308}$

2.5 Working with Complex Numbers

Addition

We add the two parts separately.

For example

$$(3 + 2i) + (4 - i) = 7 + i$$

Multiplication

We do a normal binomial multiplication.

For example:

$$(3 + 2i)(4 - 2i) = 12 - 6i + 8i - 4i^2$$
$$= 12 + 2i - 4i^2$$

But

$$i^2 = -1$$

so our final answer is $12 + 2i + 4$ or $16 + 2i$.

Division

Consider the following example:

$$\frac{3+i}{1-3i}$$

We need to find a way to remove the imaginary part from the denominator. We can do this by multiplying both numerator and denominator by $(1 + 3i)$.

$$\frac{3+i}{1-3i} = \frac{(3+i)(1+3i)}{(1-3i)(1+3i)}$$

$$= \frac{3 + 9i + i + 3i^2}{1 + 3i - 3i - 9i^2}$$

$$= \frac{3 + 10i + 3i^2}{1 - 9i^2} \qquad \text{Because} \quad i^2 = -1$$

$$= \frac{3 + 10i - 3}{10}$$

$$= \frac{10i}{10}$$

$$= i$$

2.6 Number Sets

As mentioned in the previous chapter, it is common to refer to the sets of numbers by the following letters:

- $\mathbb{N}$ The set of natural numbers
- $\mathbb{Z}$ The set of integers
- $\mathbb{Q}$ The set of rational numbers
- $\mathbb{R}$ The set of real numbers
- $\mathbb{C}$ The set of complex numbers.

We see that: $\mathbb{N} \subset \mathbb{Z} \subset \mathbb{Q} \subset \mathbb{R} \subset \mathbb{C}$.

2.7 Discrete Versus Continuous Values

Imagine turning on a tap just enough so that the tap drips. You can count the individual drops of water that come out of the tap.

Now imagine you turn the tap further, so that the water comes in a continuous stream. Now it is not possible to count anything.

Discrete values are like the drops of water. They come in separate packets, and you can count them. Natural numbers and integers are discrete. You can count how many integers there are between, say, 1 and 10.

Continuous values are like the stream of water coming from the tap. There are no separate items that you can count. Real numbers are continuous. It is not possible to say how many items there are between, say, 2.1 and 2.2.

2.8 Countable and Non-countable Sets

Consider the set of positive numbers greater than or equal to 5 and less than or equal to 10. This set would contain 6 numbers—we can count them from 5 up to 10.

But what if we put no upper limit on the set? We could still count the members—the only thing is that we would never finish counting, because the number of elements is infinite. A set like this is said to be *countable* and *infinite*.

Now consider the set of real numbers from 5 to 10. How could we count these? If we started with 5.0, then counted 5.1, what about 5.01, 5.02 etc.? And what about the numbers between 5.01 and 5.02? You can see we can never count real numbers—and in fact it can be mathematically proven that if we take any two real numbers, there is always another number in between.

The set of real numbers, $\mathbb{R}$, is *non-countable* and infinite—whereas the set of natural numbers, $\mathbb{N}$, and the set of integers, $\mathbb{Z}$, are *countable* and infinite.

So you can see that there are in fact two types of infinity—countable infinity and non-countable infinity.

2.9 Operations

In mathematics, an **operation** is a calculation from a number of input values (called **operands**) to an output value.

The basic arithmetic operations are:

- Addition (+)
- Subtraction (−)
- Multiplication (× or *)
- Division (÷ or /)

2.9.1 Commutative Operations

When we perform an operation on two terms, the operation is **commutative** if it doesn't matter which of the terms is placed first and which one is placed second. In arithmetic the following operations are commutative:

Addition

$$x + y = y + x$$

For example: $2 + 3 = 3 + 2 = 5$

Multiplication

$$x \times y = y \times x$$

For example: $2 \times 3 = 3 \times 2 = 6$

The following arithmetic operations are *not* commutative:

Subtraction

$$x + y \neq y + x$$

For example: $10 - 3 = 7$ but $3 - 10 = -7$

$$10 - 3 \neq 3 - 10$$

Division

$$x \div y \neq y \div x$$

For example: $10 \div 5 = 2$ but $5 \div 10 = 0.5$

$$10 \div 5 \neq 5 \div 10.$$

2.9.2 Associative Operations

When we perform an operation on more than two terms, the operation is **associative** if it doesn't matter how we group the terms. In arithmetic the following operations are associative:

Addition

$$x + (y + z) = (x + y) + z$$

For example: $2 + (3 + 4) = (2 + 3) + 4 = 9$

Multiplication

$$x \times (y \times z) = (x \times y) \times z$$

For example: $2 \times (3 \times 4) = (2 \times 3) \times 4 = 24$

The following arithmetic operations are *not* associative:

Subtraction

$$x - (y - z) = (x - y) - z$$

For example: $12 - (6 - 2) = 8$ but $(12 - 6) - 2 = 4$
$12 - (6 - 2) \neq (12 - 6) - 2$.

Division

$$x \div (y \div z) = (x \div y) \div z$$

For example: $12 \div (6 \div 2) = 4$ but $(12 \div 6) \div 2 = 1$

$$12 \div (6 \div 2) \neq (12 \div 6) \div 2$$

In set theory the following operations are commutative:

Union

$$A \cup B = B \cup A$$

Intersection

$$A \cap B = B \cap A$$

The following operations are *not* commutative:

Difference

$$A \backslash B \neq B \backslash A$$

Cartesian product

$$A \times B \neq B \times A$$

The following set operations are associative:

Union

$$(A \cup B) \cup C = A \cup (B \cup C)$$

Intersection

$$(A \cap B) \cap C = A \cap (B \cap C)$$

The following set operation is *not* associative:

Difference

$$(A \backslash B) \backslash C \neq A \backslash (B \backslash C)$$

2.9.2.1 Cartesian Product and Associativity

If Cartesian product is associative then:

$$(A \times B) \times C = A \times (B \times C)$$

Strictly speaking it is *not* associative, as illustrated by the following example:

Let $A = \{a, m, n\}$ $B = \{b, x, y\}$ $C = \{c, d, e, f, g\}$.

The left hand side would give us:

$$\{((a, b), c), ((a, b), d), ((a, b), e)...\}$$

The right side would give as a set such as:

$$\{(a, (b, c)), (a, (b, d)), (a, (b, e))...\}$$

However, we often assume that the Cartesian product of three sets just gives as triples.

In this case it *is* associative as both sides would gives us:

$$\{(a, b, c), (a, b, d), (a, b, e)...\}$$

2.10 Groups

A group is a **set** combined with an **operation**. For example:

The *set* of integers with the *operation* addition

However, a set and an operation do not constitute a group unless it satisfies the following criteria, all of which are explained in the subsequent sections:

1. The group contains an **identity**.
2. The group contains **inverses**.
3. The operation is **associative**.
4. The group is **closed** under the operation.

We will now explore each of these in turn.

1. **A group must contain an identity**

2.10 Groups

The set must contain an element known as the **identity** element. If the operation is applied to any element and the identity, the element will be unchanged.

For example
For the set of *integers* and *addition*, the identity is 0:

$$5 + 0 = 5 \qquad 0 + 5 = 5$$
$$7 + 0 = 7 \qquad 0 + 7 = 7$$
$$-5 + 0 = -5 \qquad -5 + 0 = -5 \quad \text{etc.}$$

- the identity element must be an element of the set;
- there is only one identity element for every group;
- the symbol that is usually used for the identity element is e.

Formally:

For a set G under the operation $*$:
 There exists an e in G, such that $a * e = a$ and $e * a = a$, for all elements a in G.

2. **A group must contain inverses**

For every element of the group, there's another element of the group such that when we use the operator on both of them, we get e, the identity.

For example
For the *integers* and *addition*, the inverse of 5 is -5 (because $5 + -5 = 0$); the inverse of -5 is 5.

- if a is the inverse of b, then it must be that b is the inverse of a;
- inverses are unique—for example, there is no other x, apart from -5, such that $5 + x = 0$.

Notation
The inverse of a is written as a^{-1}. In the above example, $a^{-1} = b$.

Formally:

For all a in G, there exists a b in G, such that $a * b = e$ and $b * a = e$.

3. **A group must be associative**

We already know the meaning of this.

For example
For the set of integers, addition is an associative operation.

Formally:

For all a, b, and c in G, $a * (b * c) = (a * b) * c$.

4. **A group must be closed under the operation**

- if there are two elements in the group, a and b, and * represents the operation, then it must be the case that $a * b$ is also in the group.
- we say that the group is **closed** under the operation.

For example
With integers and addition, the group is closed because whatever the value of a and b, $a + b$ is always an integer.

Formally:

For all a, b in G, $a * b$ is in G.

We see that the group integers with addition is formally a group because it has all of the required properties.

2.11 Notation and Terminology

You will often see a group written like this:
   $(G, *)$ where G represents the set, and * the operation.

For example:

- The set of integers with addition: $(\mathbb{Z}, +)$
- The set $\{1\}$ with multiplication: $(\{1\}, \times)$

Note

- sometimes you will see the symbol $\circ$ used instead of *;
- if a group is also commutative, it is referred to as an **abelian** group;
- a group that has associativity and closure only is called a **semigroup**;
- a semigroup that has an identity element is called a **monoid**.

Worked Example 4
Is the set $\{-1, 1\}$ under multiplication a group?

2.11 Notation and Terminology

Solution

Is there an identity element?
The identity element is 1 because $1 \times 1 = 1$ and $-1 \times 1 = -1$.

Are there inverses?
$1 \times 1 = 1$ and $-1 \times -1 = 1$, so there is an inverse for each element.

Is the operation associative?
We already know that multiplication is associative with any integers.

Is there closure?
It is closed because the results of all of the following are in the group:

$1 \times 1 \quad -1 \times -1 \quad 1 \times -1 \quad -1 \times 1$.

Therefore the set $\{-1, 1\}$ under multiplication is a group.

Worked Example 5
Is the set $\{-1, 0, 1\}$ under addition a group?

Solution

Is there an identity element?
The identity element is 0.

Are there inverses?
$-1 + 1 = 0$, $1 + -1 = 0$ and $0 + 0 = 0$, so there is an inverse for each element.

Is the operation associative?
We already know that addition is associative with any integers.

Is there closure?
It is not closed because the result of adding -1 to itself is not in the group – neither is the result of adding 1 to itself.

Therefore the set $\{-1, 0, 1\}$ under addition is not a group.

Worked Example 6
Is the set of integers under multiplication a group?

Solution

Is there an identity element?
The identity element is 1.

Are there inverses?
If you take an integer such as 7, there is no integer that will result in 1 if multiplied by 7. This applies to all integers except 1. So it does not satisfy the inverse property.

The set of integers under multiplication is therefore not a group.

Worked Example 7
Is the set of rational numbers excluding zero under multiplication a group?

Solution

Is there an identity element?
The identity element is 1.

Are there inverses?
We can find an inverse for every element.

For example: The inverse of $\frac{5}{7}$ is $\frac{7}{5}$ because $\frac{5}{7} \times \frac{7}{5} = 1$.
The same applies to every rational number (excluding zero).

Is the operation associative?
We already know that multiplication is associative with all numbers.

Is there closure?
It is closed because the result of any possible multiplication always results in another rational number.

$$\frac{a}{b} \times \frac{b}{d} = \frac{ac}{bd}$$

Therefore the set of rational numbers excluding zero under multiplication is a group. But notice that if we include zero, it is not a group because there is no inverse for zero.

2.12 Relevance to Computing

Number theory and group theory are essential parts of the foundations of mathematics. But for students studying computing they are essential for areas such as algorithm design, coding theory, and encryption systems.

2.13 Exercises

1. Use the laws of set algebra to simplify the following expression:

$$A \cup (\overline{A} \cap B)$$

2. Use set algebra to show that:

$$B \cap \overline{(A \cap B)} = B \cap \overline{A}$$

3. In the last chapter you drew Venn diagrams to prove the following:

$$A \backslash B = A \cap \overline{B}$$

   (a) Bearing this in mind, use set algebra to show that:

$$B \cup (A \backslash B) = B \cup A$$

   (b) Verify that this is true by drawing a Venn diagram.
4. Consider the following numbers: 2.25   –335   2   – 3.75   0
     Which of these numbers are:
     (a) Real numbers   (b) Integers   (c) Natural numbers?
5. Which of the following is a rational number? Give a reason for your answer.
     7.5 e (Euler's constant) $\sqrt{7}$ $\sqrt{25}$
6. Simplify the following expressions containing complex numbers:
     (a) $i(3 + i)$ (b) $(2 - 3i)(4 + i)$ (c) $(3 - i)^2$ (d) $(1 + i)^3$
     (e) $\frac{3+4i}{1+i}$
7. Consider the following sets:

   A = the set of integers less than 10
   B = the set of natural numbers less than 10
   S = the set of people living in London
   T = the set of real numbers greater than 1000

   In each case, state whether the set is:
   (a) countable or non-countable
   (b) finite or infinite.

8. Consider the following combinations of a set and an operation, and state, giving reasons, which of them constitute a formal group:
   (a) Natural numbers under subtraction
   (b) The set {0} under addition
   (c) The set {1} under multiplication
9. Show that the set of natural numbers starting from 0 ($\mathbb{N}_0$) under addition is a monoid but not a group.
10. Show that the set of natural numbers starting from 1 ($\mathbb{N}_1$) under addition is a semigroup but not a group.

Relations 3

At the end of this chapter you should be able to:

- explain the meaning of the term **relation**;
- represent a **binary relation** pictorially and as a set of ordered pairs;
- find the **inverse** of a given relation;
- determine whether a particular relation on a set is an **equivalence relation**.

3.1 Introduction

In this chapter we are going to deal with the topic of **relations**, which is to do with the ways in which elements of different sets can be related to each other. This subject forms the basis upon which other very important mathematical concepts such as **functions** are derived, and also has a direct application in computing, since it is the theoretical base that underlies the structure of **relational databases**.

3.2 Relations: Definition and Notation

A **relation** is a set of connections from one set to another set.

Consider a small college that specialises in science. Four subjects are taught – physics, chemistry, biology and mathematics. There are three members of the teaching staff – Juliet, Ade and Meena. We can define two sets, the set of teachers and the set of subjects. We will call them T and S respectively:

$$T = \{\text{JULIET, ADE, MEENA}\}$$

Fig. 3.1 A relation between teachers and subjects

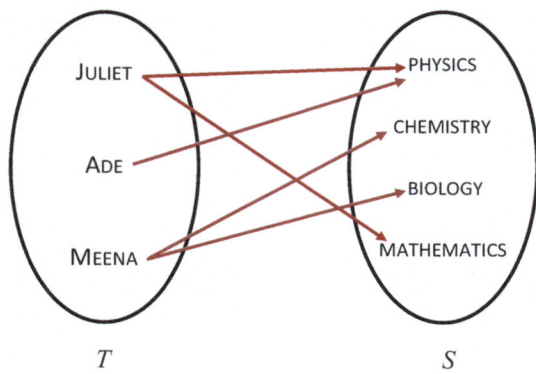

$S = \{\text{PHYSICS, CHEMISTRY, BIOLOGY, MATHEMATICS}\}$

We need to consider which teachers teach which subjects. One way we can represent this is by means of a diagram, as show in in Fig. 3.1.

The diagram shows a **relation** between the set of teachers and the set of subjects. We often use R to represent a relation, and express R as a *set of ordered pairs*. In our example:

$R = \{$(JULIET, PHYSICS), (JULIET, MATHEMATICS), (ADE, PHYSICS), (MEENA, CHEMISTRY), (MEENA, BIOLOGY)$\}$

We can express the relationship between the individual pairs as follows:

JULIET $\mathcal{R}$ MATHEMATICS

We read this as: JULIET *is related to* MATHEMATICS
In our case it means: JULIET *teaches* MATHEMATICS

Worked Example 1

The diagram in Fig. 3.2 represents the relation "*is produced by*" between the set F (food) and the set A (animals):

(a) Represent the relation in terms of a set of ordered pairs.
(b) Write in words: MILK $\mathcal{R}$ GOATS

Solution

(a) $R = \{$ (MILK, COWS), (MILK, GOATS), (EGGS, HENS) $\}$
(b) MILK *is produced by* GOATS

3.2 Relations: Definition and Notation

Fig. 3.2 The relation *is-produced-by*

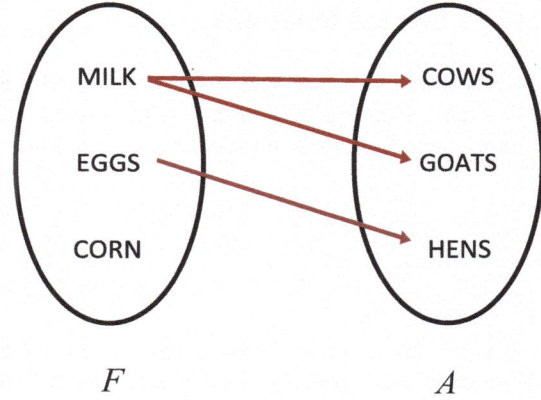

Worked Example 2

A and B are two sets and R is a relation from set A to set B.

$$A = \{a, b, c, d\}$$

$$B = \{x, y, z\}$$

$$R = \{(a, x), (b, y), (c, y), (d, x)\}$$

Represent the relation R pictorially.

Solution
See Fig. 3.3

Fig. 3.3 Solution to worked example 2

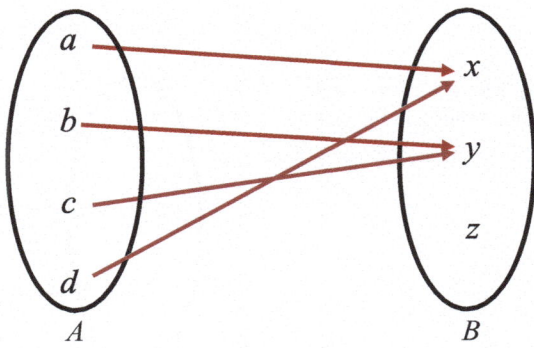

3.3 Inverse Relations

Consider the relation, R, from a set A to a set B, shown in Fig. 3.4

The relation in Fig. 3.5 represents the **inverse** of this relation, written as R^{-1}. Expressing R as a set of ordered pairs, we have:

$$R = \{(a, x), (b, y), (c, y), (d, z)\}$$

$$R^{-1} = \{(x, a), (y, b), (y, c), (z, d)\}$$

In general, if R is a relation from a set A to set B, then the inverse, R^{-1}, is a relation from a set B to a set A and is found by reversing the pairs.

We can use the notation for set comprehension that we learnt previously to give a formal definition of the inverse:

$$R^{-1} = \{(b, a) | (a, b) \in R\}$$

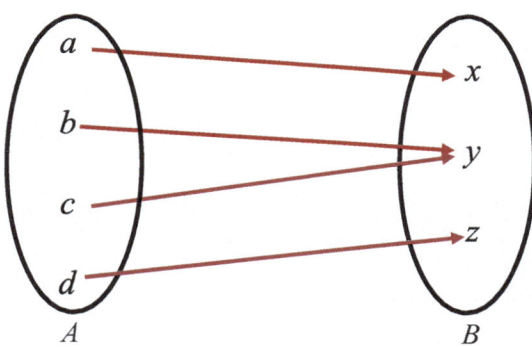

Fig. 3.4 A relation from A to B

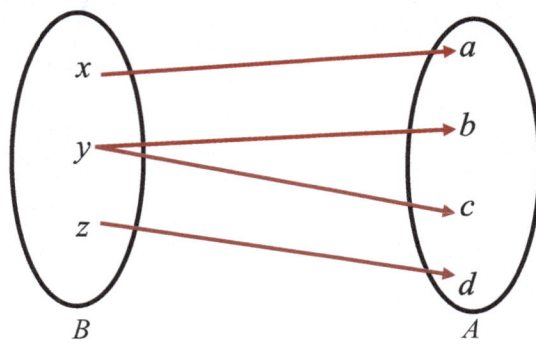

Fig. 3.5 The inverse of the relation shown in Fig. 3.4

3.6 Symmetric Relations

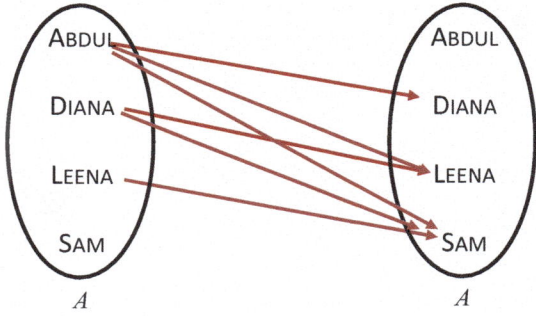

Fig. 3.6 The relation "is older than"

3.4 Relations on a Set

When a relation is between two identical sets, we talk about a relation *on a set*. For example consider the following set, A, consisting of four people.

$$A = \{\text{ABDUL, DIANA, LEENA, SAM}\}$$

Now consider the relation R, *is older than*, shown in Fig. 3.6.
Writing the relation as a set of ordered pairs gives:

$$R = \{(\text{ABDUL,DIANA}), (\text{ABDUL, LEENA}), (\text{ABDUL, SAM}),$$
$$(\text{DIANA, LEENA}), (\text{DIANA, SAM}), (\text{LEENA, SAM})\}$$

3.5 Digraphs of Relations

In Chap. 10 you will study graph theory. There you will see that a graph is a diagram that consists of a set of vertices and a set of connections - in a directed graph, or **digraph**, those connections have a direction. A digraph is a useful way of representing a relation on a set.

The digraph in Fig. 3.7 represents the relation $\{(a, b), (b, b), (a, c), (a, d), (b, c), (b, d), (c, d)\}$ on the set $\{a, b, c, d\}$.

3.6 Symmetric Relations

Consider the set, A:

$$A = \{1, 2, 3, 4\}$$

Now consider the following relation, R, on this set:

$$R = \{(1, 2), (2, 1), (2, 3), (3, 2)(3, 4), (4, 3)\}$$

Fig. 3.7 A digraph of a relation

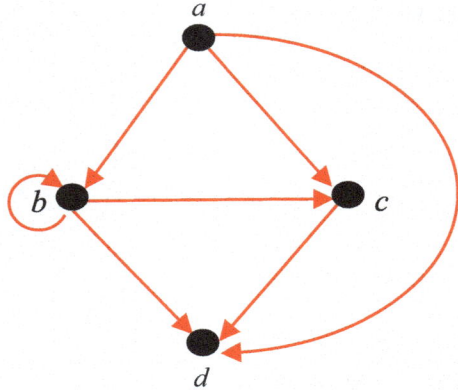

We can see that:
1 $\mathcal{R}$ 2 and 2 $\mathcal{R}$ 1.
2 $\mathcal{R}$ 3 and 3 $\mathcal{R}$ 2.
3 $\mathcal{R}$ 4 and 4 $\mathcal{R}$ 3.

This is an example of a **symmetric** relation. In such a relation, for every pair (a, b) there is also a pair (b, a).

Put another way, whenever $a\ \mathcal{R}\ b$ then we also find that $b\ \mathcal{R}\ a$.

We can see that our previous example *is older than* is <u>not</u> symmetric.

For example: ABDUL *is older than* DIANA, but DIANA *is not older than* ABDUL.

3.7 Reflexive Relations

Consider again the set, A

$$A = \{1, 2, 3, 4\}$$

Now consider the following relation, R, on this set:

$$R = \{(1, 1),\ (2, 1),\ (2, 2),\ (2, 4)\ (3, 3),\ (4, 4)\}$$

We can see that:
1 $\mathcal{R}$ 1.
2 $\mathcal{R}$ 2
3 $\mathcal{R}$ 3
4 $\mathcal{R}$ 4

This is an example of a **reflexive** relation. In such a relation, every element is related to itself.

Put another way, for every element, *a*, we find that $a \mathcal{R} a$. We can see that our previous example *is older than* is <u>not</u> reflexive.

For example ABDUL *is not older than* ABDUL, and so on.

3.8 Transitive Relations

Once again, consider the set, *A*

$$A = \{1, 2, 3, 4\}$$

Now consider the following relation, *R*, on this set:

$$R = \{(1, 2), (1, 3), (1, 4), (2, 3)(2, 4), (3, 4)\}$$

We can see that:
1 $\mathcal{R}$ 2, 2 $\mathcal{R}$ 3 and 1 $\mathcal{R}$ 3.
1 $\mathcal{R}$ 3, 3 $\mathcal{R}$ 4 and 1 $\mathcal{R}$ 4.
2 $\mathcal{R}$ 3, 3 $\mathcal{R}$ 4 and 2 $\mathcal{R}$ 4.

This is an example of a **transitive** relation. In such a relation, if $a \mathcal{R} b$ and $b \mathcal{R} c$, then $a \mathcal{R} c$.

We can see that our previous example *is older than* <u>is</u> transitive.
For example:

ABDUL *is older than* DIANA, DIANA *is older than* LEENA, and ABDUL *is older than* LEENA.
ABDUL *is older than* LEENA, LEENA *is older than* SAM, and ABDUL *is older than* SAM.
DIANA *is older than* LEENA, LEENA *is older than* SAM, and DIANA *is older than* SAM.

3.9 Equivalence Relations

If a relation is symmetric, reflexive and transitive then it is described as an **equivalence** relation.

Worked Example 3
Consider the relation *is greater than or equal to* ($\geq$) on the set of integers. Is this an equivalence relation?

<u>Solution</u>
It is a reflexive relation, because for every number, *a*, $a \geq a$.
  For example $1 \geq 1$, $3 \geq 3$ and so on.
  It is a transitive relation, because whenever $a \geq b$ and $b \geq c$, then $a \geq c$.

For example $7 \geq 6$, $6 \geq 4$ and $7 \geq 4$.

However, it is *not* a symmetric relation, because it is *not* true that whenever $a \geq b$ then $b \geq a$.

For example $7 \geq 6$ but $6 \geq 7$.

Is greater than or equal to is therefore *not* an equivalence relation.

3.10 Equivalence Classes

Consider the set, A

$$A = \{1, 2, 3, 4\}$$

Now consider the following equivalence relation, R, on this set:

$$R = \{(1, 1), (2, 2), (3, 3), (4, 4), (1, 2), (2, 1), (2, 3), (3, 2), (1, 3), (3, 1)\}$$

For each element of the set there is an **equivalence class**, defined as the set of elements that that element is related to.

It is written with square brackets:

So : $[1] = \{1, 2, 3\}$ $[2] = \{1, 2, 3\}$ $[3] = \{1, 2, 3\}$ $[4] = \{4\}$.

Notice here that we have arrived at two disjoint sets $\{1, 2, 3\}$ and $\{4\}$.

Equivalence classes always partition the set. In other words, any two equivalence classes are either equal or disjoint.

Formally, given a set S and an equivalence relation R on S, the *equivalence class* of an element a in S is the set:

$$\{x \in S \mid x \mathcal{R} a\}$$

Worked Example 4

Consider the set, A

$$A = \{a, b, c\}$$

Now consider the following equivalence relation, R, on this set:

$$R = \{(a, a), (a, b), (b, a), (b, b), (c, c)\}$$

Find the equivalence classes for this relation.

Solution

$$[a] = a, b \quad [b] = a, b \quad [c] = c.$$

3.11 Binary Relations

The relations we have just seen are **binary** relations: they are relations between *two* sets. In our first worked example we had:

$$F = \{\text{MILK, EGGS, CORN}\}$$

$$A = \{\text{COWS, GOATS, HENS}\}$$

and we looked at the particular relation:

$$R = \{\,(\text{MILK, COWS}), (\text{MILK, GOATS}), (\text{EGGS, HENS})\,\}$$

In a binary relation, all the possible relations that could exist are given by the Cartesian product of the two sets.

In this case:

$$F \times A = \begin{array}{l} \{(\text{MILK,COWS}), (\text{MILK, GOATS}), (\text{MILK, HENS}), \\ (\text{EGGS, COWS}), (\text{EGGS, GOATS}), (\text{EGGS, HENS}), \\ (\text{CORN,COWS}), (\text{CORN, GOATS}), (\text{CORN, HENS})\} \end{array}$$

A particular relation, R, is a subset of the Cartesian product.

3.12 *n*-Ary Relations

We saw that in our previous example we had two sets:

$$Food = \{\text{MILK, EGGS, CORN}\}$$
$$Animals = \{\text{COWS, GOATS, HENS}\}$$

Let us add another set into the mix:

$$Farms = \{\text{MANOR FARM, CITY FARM}\}$$

Imagine that they keep goats and hens at Manor Farm and cows and hens at City Farm. A relation that shows which animal produces which food at which farm would look like this:

$$R = \{(MILK, GOATS, MANOR FARM), \\ (MILK, COWS, CITY FARM), \\ (EGGS, HENS, MANOR FARM), \\ (EGGS, HENS, CITY FARM)\}$$

A relation like this, which relates three sets together, is called a **ternary** relation. In general we can have 2-ary, 3-ary, 4-ary relations and so on.

3.13 Relevance to Computing: Relational Databases

Databases are of enormous importance to computing – and nowadays most databases are based on the *relational* model.

Relational databases use tables to store information. Each row of the table is called a **record** and each column is called a **field**. An example is shown in Table 3.1.

Each row of the table is one element in a *relation* (in this case a 5-ary relation). A typical database consists of many related tables: Table 3.2 is connected to Table 3.1 by the *Department_No* field.

Relational databases are created and queried by a special language based on the theory of relations. This is known as **Structured Query Language** (SQL).

Table 3.1 A table in a relational database storing employee information

Employee_No	Name	Department_No	Date_of_Birth	Salary
108765	Patty O'Furniture	3	14.09.1960	43,000
282098	Isadore open	1	30.08.1981	28,000
291073	Justin case	2	28.02.1975	55,000
365498	Anne T body	4	31.10.1990	22,000

Table 3.2 A second table in the database

Department_No	Department_Name
1	Accounts
2	Sales
3	Human resources
4	Marketing

Fig. 3.8 The relation *plays-for*

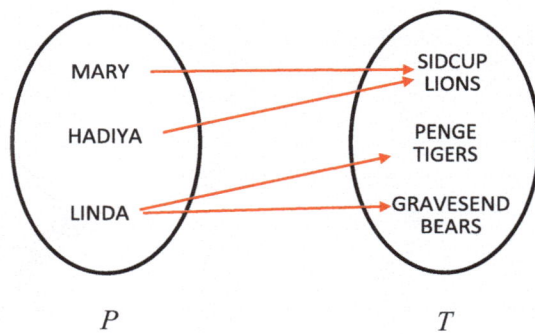

3.14 Exercises

1. The diagram in figure 3.8 represents the relation *"plays for"* between a set of players, P, and a set of teams, T:
   (a) Represent the relation in terms of a set of ordered pairs.
   (b) Write in words: Mary $\mathcal{R}$ Sidcup Lions
2. A and B are two sets and R is a relation from set A to set B, where:

$$A = \{1, 2, 3\}$$
$$B = \{x, y\}$$
$$R = \{(1, x), (2, x), (3, y)\}$$

   Represent the relation R pictorially.

3. A relation R is specified as follows:

$$R = \{(a, 2), (d, 9), (b, 4), (c, 7), (a, 1)\}$$

   Give the value of R^{-1}, the inverse of this relation.

4. Consider the relation *"is less than"* on the set of integers.
      State whether this relation is:
   (a) Symmetric
   (b) Reflexive
   (c) Transitive

Fig. 3.9 A relation on the set $\{a, b, c, d\}$

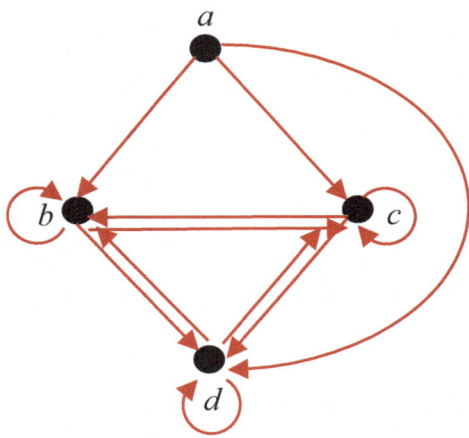

In each case give a reason for your answer.

5. Consider the digraph in Fig. 3.9 which shows a relation on the set $\{a, b, c, d\}$.

State whether this relation is:

   (a) Reflexive
   (b) Symmetric
   (c) Transitive

Functions

At the end of this chapter you should be able to:

- define the term **function**, and determine whether a particular binary relation is or is not a function;
- **apply** a function to a particular input;
- write an appropriate **function signature**;
- where appropriate, specify a function as a mathematical formula;
- describe how a function can effectively have multiple inputs;
- explain the term **function composition**.
- determine whether a function is **injective**, **surjective** or **bijective**.

4.1   Introduction

In this chapter we cover the very important mathematical concept of a function. As you will find out, a function is a type of relation, and an understanding of functions is vital not only to the area of mathematics but also to the world of computer science.

4.2   Definition

A **function** is a special sort of binary relation. It is a relation in which each element of the first set relates to one and only one element in the second set. This means we can assign more than one element of the first set to the same element of the second set, but never the other way round. It also means that every element of the

Fig. 4.1 Functions and relations

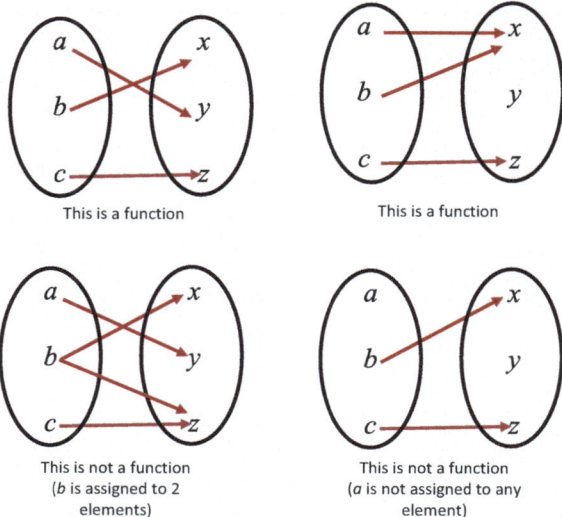

first set must be assigned to a member of the second set. The examples in Fig. 4.1 will make this clear:

The first set is known as the **domain** of the function, the second set is known as the **codomain** (or **range**).

The element in the second set that corresponds to a particular element of the first set is called the **image** of that element. In the first diagram, for example, y is the image of a. In a function, every element of the domain must have *one and only one* image in the codomain.

Worked Example 1
Which of the diagrams in Fig. 4.2 represents a function?

Solution

(a) is not a function – b has no image in the codomain.
(b) is a function.
(c) is not a function – c maps to two elements in the codomain.
(d) is a function.

4.3 Functions as Mappings

We see that a function from set A to set B is a set of assignments from A to B. It can also be described as a *mapping* from A to B. In fact it is a *many-to-one* mapping, because many elements of A can map to one element of B.

4.3 Functions as Mappings

Fig. 4.2 Worked example 1

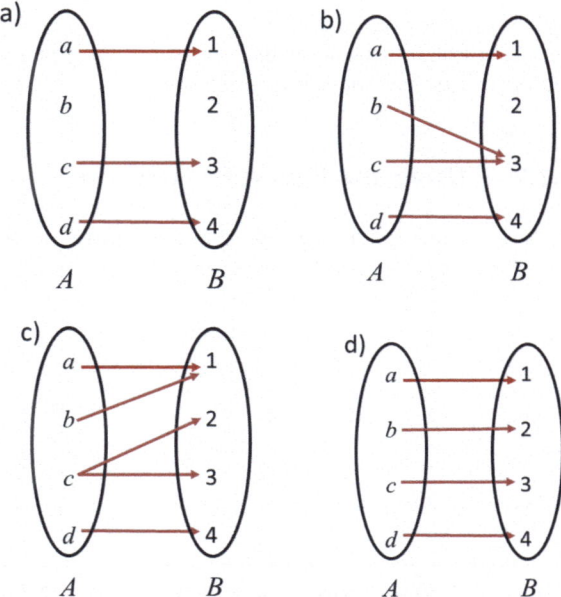

4.3.1 Practical Example

On a computer network we need to keep records of users of the system and their corresponding passwords.

We would need a function that maps names of users (N) to passwords (P). This is represented in Fig. 4.3.

Fig. 4.3 A function representing users of a network and their passwords

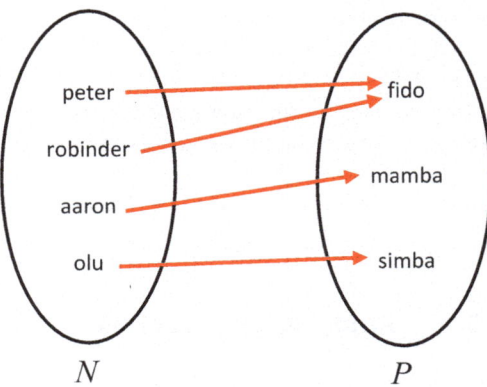

Note
This is a function because more than one user can have the same password, but each user can have only one password.

4.3.2 Using the Password Function

Let's name our function f. We can now **apply** our function to each element of N in order to find the user's password. We write this as follows:

$$f(peter) = fido$$
$$f(robinder) = fido$$
$$f(aaron) = mamba$$
$$f(olu) = simba$$

We pronounce this as f of *peter*, f of *robinder* and so on.

4.4 A Function as an Input/Output Device

A function is effectively a device that transforms an input to an output. Our password example, for instance, took a username as an input and transformed it into a password as shown in Fig. 4.4.

4.5 Functions as Formulae

When a function takes a number and outputs another number, it can often be specified as a mathematical formula.

Imagine a function, f, that behaves like this for three different inputs as shown in Fig. 4.5.

In this case, the function takes an input and adds 5 to it. If the function behaves like this for every input, we can express our function as a formula:

$$f(x) = x + 5$$

4.5.1 Applying our Function

Now that we have our function expressed like this: $f(x) = x + 5$, we can apply it to any input. For example:

$$f(2) = 7$$
$$f(0) = 5$$
$$f(-2) = 3$$

4.5 Functions as Formulae

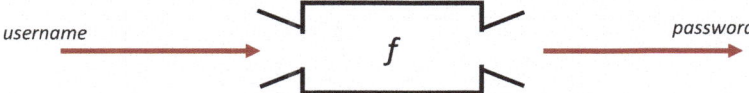

For each individual input, we get the following outputs:

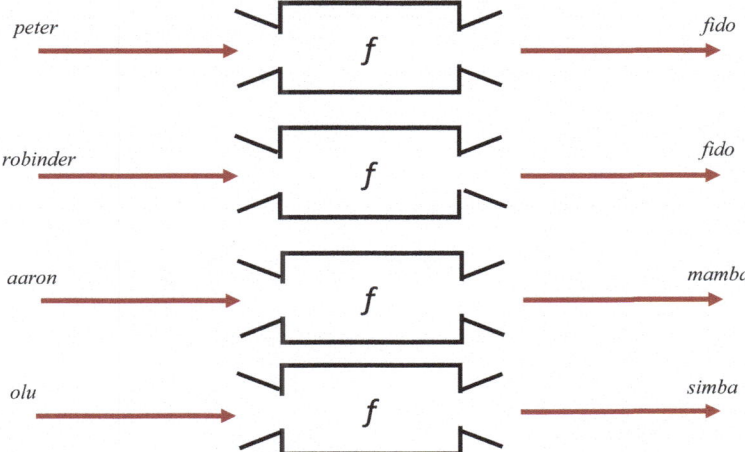

Fig. 4.4 A function as an input/output device

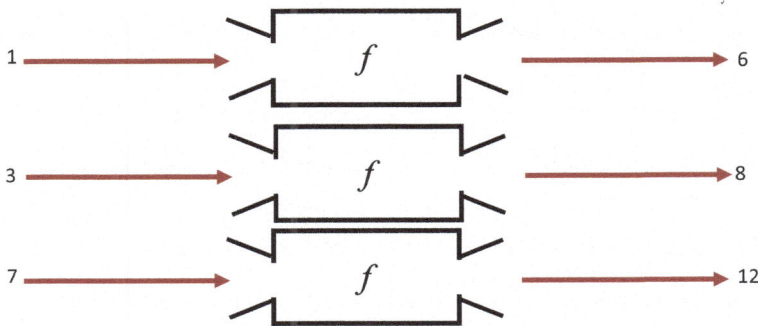

Fig. 4.5 Possible input/output pairs for a particular function, f

4.5.2 Another Example

Imagine a function that always outputs the square of the input. We can write this as:

$$f(x) = x^2$$

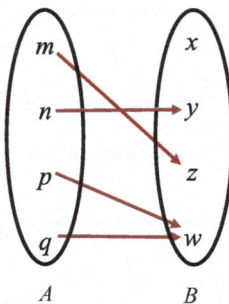

Fig. 4.6 Worked example 2

Some sample inputs would give us:

$$f(2) = 4$$
$$f(3) = 9$$
$$f(-2) = 4$$

Worked Example 2
A function f, which maps from a set A to a set B, is represented pictorially in Fig. 4.6.
What is the value of the following?

(a) $f(m)$ (b) $f(n)$ (c) $f(p)$ (d) $f(q)$

Solution

(a) $f(m) = z$ (b) $f(n) = y$ (c) $f(p) = w$ (d) $f(q) = w$

Worked Example 3
A function f is specified as follows:

$$f(x) = 2x - 1$$

What is the value of the following?

(a) $f(3)$ (b) $f(-2)$ (c) $f(0)$

Solution

(a) $f(3) = 2 \times 3 - 1$

4.5 Functions as Formulae

$$= 6-1$$
$$= 5.$$

(b) $f(-2) = 2 \times (-2) - 1$
$$= -4-1$$
$$= -5$$

(c) $f(0) = 2 \times 0 - 1$
$$= 0-1$$
$$= -1$$

Worked Example 4
A function f is specified as follows:

$$f(x) = 3x^2 - 10$$

What is the value of the following?

(a) $f(5)$ (b) $f(-4)$ (c) $f(0)$

Solution

(a) $f(5) = 3 \times 5^2 - 10$
$$= 3 \times 25 - 10$$
$$= 75 - 10$$
$$= 65$$

(b) $f(-4) = 3 \times (-4)^2 - 10$
$$= 3 \times 16 - 10$$
$$= 48 - 10$$
$$= 38$$

(c) $f(0) = 3 \times 0 - 10$
$$= 0 - 10$$
$$= -10$$

4.6 Functions as Equations

In mathematics we often designate a letter (such as y) to a function, and write the function as an equation.

Consider the function:

$$f(x) = 3x + 1$$

If we let $f(x) = y$, then we have the equation: $y = 3x + 1$.

4.7 The Signature of a Function

We usually give a function a name. In the password example we called our function f.

When we specify a function, we first need to provide its **signature**:

$$f : Usernames \to Passwords$$

This means that the function f maps from the set of *Usernames* to the set of *Passwords*. This is sometimes read as f maps *Usernames* into *Passwords*.

4.8 Specifying a Function

When we specify a function, we need to do two things: provide its signature, and say what it does. In general, if f is a function from set A to set B we write its signature as:

$$f : A \to B$$

If an element x in the domain maps onto an element y in the codomain, we write:

$$f(x) = y$$

As we saw earlier, we pronounce $f(x)$ as f of x. An example of the above function could be that shown in Fig. 4.7:

Applying the function to each element of A gives us:

$$f(a) = 1$$

$$f(b) = 3$$

$$f(c) = 3$$

This, together with the signature, fully specifies the function.

Fig. 4.7 An example function

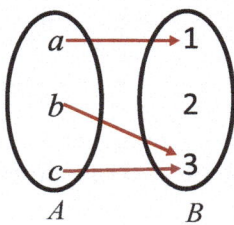

4.9 Specifying Formulaic Functions

Consider one of the previous examples, $f(x) = x^2$.

Consider the following inputs:

$$f(3) = 9$$

$$f(4) = 16$$

$$f(-2) = 4$$

$$f(-3) = 9$$

The input is an integer, but the output is always a natural number (we never produce a negative number by taking the square).

We could write our function like this:

$$f : \mathbb{Z} \to \mathbb{N}$$

$$f(x) = x^2$$

When we specify functions which are formulae, the signature will involve number sets such as the above. Often, when working with formulae and equations, we assume that - unless otherwise stated - we are dealing with real numbers; in other words the function maps from $\mathbb{R}$ to $\mathbb{R}$.

4.10 Not all Mappings are Functions

Consider the following mapping:

$$f : \mathbb{N} \to \mathbb{R}$$

$$f = \sqrt{x}$$

A number in the domain such as 4 will have two images in the codomain, namely 2 and -2.

So f is not a function.

4.11 Functions with More than One Input

In computer graphics, images are formed from pixels made up of three primary colours - red, green and blue. Mixing these can produce three secondary colours - cyan, magenta, yellow:

$$CYAN = BLUE + GREEN$$

$$MAGENTA = RED + BLUE$$

$$YELLOW = RED + GREEN$$

We can define two sets, P to represent primary colours and S to represent secondary colours:

$$P = \{RED, GREEN, BLUE\}$$

$$S = \{CYAN, MAGENTA, YELLOW\}$$

A function that produces secondary colours from primary colours would look like that shown in Fig. 4.8.

The first set is actually the Cartesian product of two sets. In this case the two sets are the same, P and P – but they could be two different sets.

If we call the above function f, then we see that the signature of f is:

$$f : P \times P \rightarrow S$$

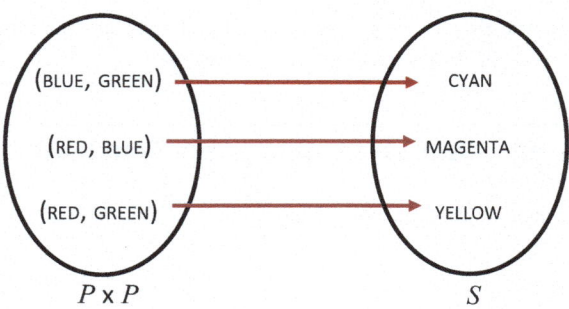

Fig. 4.8 A function that produces secondary colours from primary colours

What we have done here is effectively to extend the notion of a function to be able to accept more than one input.

When applying the function we should write, for example:

$$f((\text{RED}, \text{BLUE})) = \text{MAGENTA}$$

However, we normally drop the double brackets and write:

$$f(\text{RED}, \text{BLUE}) = \text{MAGENTA}$$

In general:

$$f(x, y) = z$$

Worked Example 5
Consider the following function:

$$f : \mathbb{R} \times \mathbb{R} \to \mathbb{R}$$

$$f(x, y) = x^2 + y - 1$$

State the value of:

(a) $f(1, 2)$ (b) $f(0, 0)$

Solution

(a) $f(1, 2) = 1^2 + 2 - 1 = 2$

(b) $f(0, 0) = 0^2 + 0 - 1 = -1$

4.12 Function Composition

Function composition means applying one function to the result of another function. Consider two functions:

$$g(x) \text{ and } f(x)$$

$$f(g(x)) \text{ is the composition of } f \text{ and } g.$$

It means apply g, then apply f to the result.

Fig. 4.9 An injective function

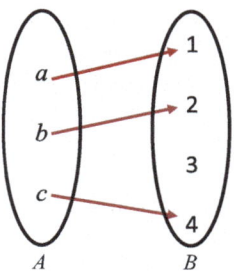

This can also be written as: $(f \circ g)(x)$.

Worked Example 6

$$g(x) = 2x + 3 \quad f(x) = x^2$$

Calculate:

$$f(g(2))$$

Solution

$$g(2) = 2 \times 2 + 3 = 7$$

$$f(7) = 7^2 = 49$$

4.13 Injective Functions

A function from A to B is **injective** (also known as **one-to-one**) if different elements in the domain all have distinct images, as shown in the example in Fig. 4.9.

4.14 Surjective Functions

A function from A to B is **surjective** (also known as **onto**) if each element of B is an image of some element of A – see Fig. 4.10.

4.15 Bijective Functions

A function which is both injective and surjective is described as **bijective**, as shown in Fig. 4.11.

Fig. 4.10 A surjective function

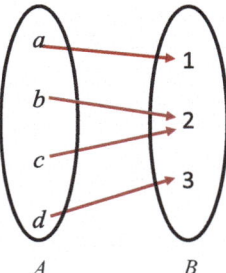

Fig. 4.11 A bijective function

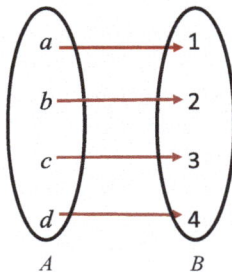

4.16 Application to Computing

4.16.1 Computer Programming

Every programming language provides a means of grouping lines of code together to perform a particular task. Such routines are given different names in different languages—in Java for example they are called *methods*; in Python they are called *functions*.

A method or function can accept multiple inputs and optionally can output a value. Any method that outputs a value is equivalent to a function in mathematics.

Consider a mathematical function (which we will call *myFunction*) that is specified as follows:

$$myFunction : f : \mathbb{Z} \times \mathbb{Z} \to \mathbb{Z}$$

$$myFunction(x, y) = 2x + y$$

In Java, for example, we would code a corresponding method as follows:

```
int myFunction(int firstIn, int secondIn)
{
    return 2 * firstIn + secondIn;
}
```

In Python the same function would look like this:

$$def\ myFunction(firstIn, secondIn):$$
$$return\ 2 * firstIn + secondIn$$

4.16.2 Business Software

In everyday life we often need to do such things as adding several numbers together or finding an average. This is particularly true in the world of business – and therefore business software such as databases and spreadsheets come packed with built-in functions to perform tasks such as these. A commercial application such as Microsoft Excel^{TM} contains literally hundreds of pre- defined functions that accept a number of inputs (often as cells in the spreadsheet) and output the result.

In Excel^{TM}, such functions include *sum, average, max, min* and many, many more.

To find the average of the numbers in the cell range A1:A4, for example, we would enter:

$$= AVERAGE(A1:A4).$$

4.17 Exercises

1. Which of the diagrams in Fig. 4.12 represents a function?
2. A function f, which maps from a set A to a set B, is represented pictorially in Fig. 4.13:
   What is the value of the following?

   (a) $f(u)$ (b) $f(v)$ (c) $f(w)$ (d) $f(x)$

3. A function f is specified as follows:

$$f : \mathbb{Z} \times \mathbb{Z} \to \mathbb{Z}$$

$$f(x) = 4x^2 - 5$$

   What is the value of the following?

   (a) $f(3)$ (b) $f(-1)$ (c) $f(0)$

4.17 Exercises

Fig. 4.12 Exercise 1

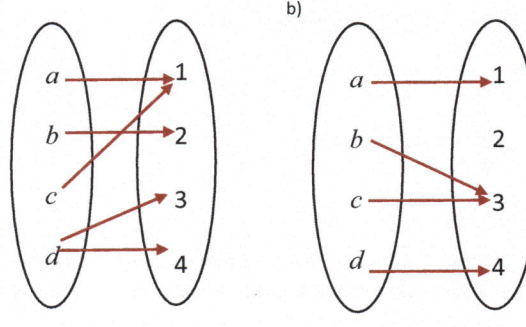

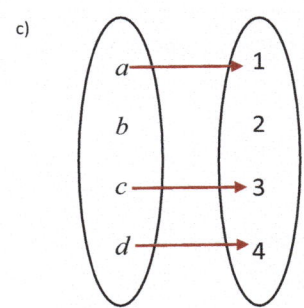

Fig. 4.13 Exercise 2

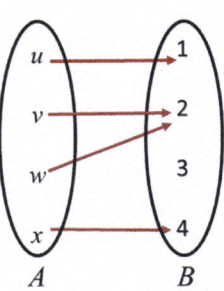

4. Consider the following function:

$$f : \mathbb{R} \times \mathbb{R} \to \mathbb{R}$$

$$f(x, y) = 2x^2 + 3y$$

State the value of:

(a) $f(2, 0)$ (b) $f(1, -1)$

5. Consider the following functions:

$$g(x) = 3x + 1 \qquad f(x) = x^3$$

Calculate:

$$f(g(3))$$

6. Write a complete specification (signature and behaviour) of a function that accepts two integers and outputs a number which is twice the sum of these two integers.
7. Consider the functions shown in Fig. 4.14. For each one, say whether it is an injective function, a surjective function, neither or both.
8. Consider the following function:

$$f : \mathbb{R} \to \mathbb{R}$$

$$f(x) = x^2$$

Is this function

        (a) surjective?   (b) injective?

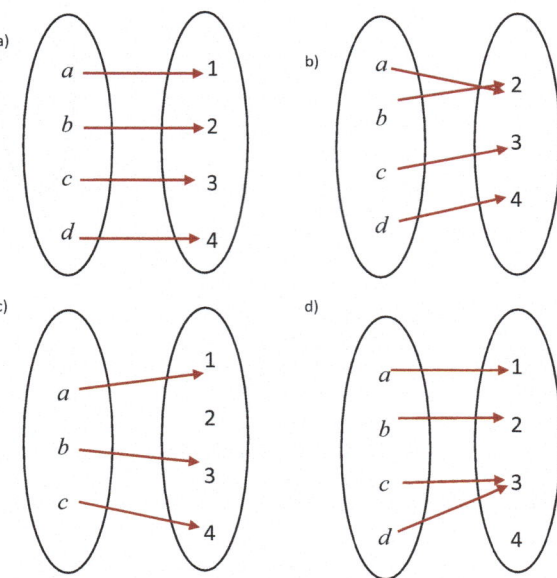

Fig. 4.14 Exercise 7

Propositional Logic

At the end of this chapter you should be able to:

- define the term **proposition**;
- provide truth tables for the simple logical operators, including NOT, AND, OR, IMPLICATION, EQUIVALENCE and EXCLUSIVE OR;
- explain and interpret the above truth tables;
- construct truth tables for compound expressions, and determine whether two expressions are logically equivalent;
- explain the terms **tautology** and **contradiction**;
- use the laws of propositional logic to perform simple algebraic operations on logical expressions;
- provide examples of how mathematical logic is applied to the field of computing.

5.1 Introduction

In this chapter and the next we are moving on to a new topic, mathematical logic, which is a well-established branch of mathematics, and is essential for the understanding of a many important topics in computer science, especially in programming. This chapter concentrates on **propositional logic**, which involves performing operations on simple statements which can be **true** or **false**. We will study the use of **truth tables** that define rules for these operations, and learn some simple laws that enable us to develop an algebra of propositions.

5.2 Mathematical Logic

In mathematical logic we try to place a rigorous mathematical framework around everyday natural language. There are two main branches of logic that we will study in this book:

- **propositional logic (this chapter)**
- **predicate logic (next chapter)**

5.3 Propositions

A **proposition** is a statement that can be either *true* or *false*. We say that it has a **truth value**.

Examples of propositions that have a value of true:

- The angles of a triangle add up to 180 degrees.
- Paris is the capital of France.
- $3 + 2 = 5$.

Examples of propositions that have a value of false:

- The angles of a triangle add up to 360 degrees.
- Paris is the capital of Scotland.
- $3 + 2 = 7$

We can represent propositions by variable names such as P or Q. For example:
 P: It is sunny.
 Q: Today is Tuesday.

5.4 Logical Operators (Connectives)

We can join two simple propositions together to form a compound statement, using **operators** (or **connectives**) that try to capture the meaning of simple words like "and" and "or". The truth value of a compound statement will depend on the value of each of the two simple statements.

Each connective has a set of rules that are defined in the form of a **truth table**.

5.4.1 The AND operator ($\wedge$)

If P and Q are propositions then:
 $P \wedge Q$ means: P AND Q.

5.4 Logical Operators (Connectives)

Fig. 5.1 The truth table for AND

P	Q	P∧Q
T	T	T
T	F	F
F	T	F
F	F	F

The meaning of the AND operator is defined in the truth table shown in Fig. 5.1 (where T means true and F means false).

If P is the statement "It is Wednesday", and Q is the statement "It is February", then $P \wedge Q$ is the statement "It is Wednesday and it is February".

The compound statement is true only if both of the simple statements are true – otherwise it is false.

This is also known as **conjunction**.

Worked Example 1

If P and Q represent the following statements:

P: 102 < 50.

Q: Nigeria is in Africa.

What is the value of $P \wedge Q$?

Solution

P is false.
Q is true.

Looking at the third line of the truth table we see that F ∧ T is false.
Therefore, $P \wedge Q$ is false.

5.4.2 The OR Operator (∨)

If P and Q are propositions then:

$P \vee Q$ means: P OR Q.

The meaning of the OR operator is defined in the truth table in Fig. 5.2.

If P is the statement "It is Wednesday", and Q is the statement "It is February", then $P \vee Q$ is the statement "It is Wednesday or it is February".

Fig. 5.2 The truth table for OR

P	Q	P∨Q
T	T	T
T	F	T
F	T	T
F	F	F

The compound statement is true if either one of the simple statements is true – it is false only when both statements are false.

This is also known as **disjunction**.

Worked Example 2
If P and Q represent the following statements:

P: 102 < 50.
Q: Nigeria is in Africa.

What is the value of $P \vee Q$?

Solution
P is false.

Q is true.

Looking at the third line of the truth table we see: $F \vee T$ is true.

Therefore, $P \vee Q$ is true.

5.4.3 The NOT Operator (¬)

Unlike the previous two operators, the **not** operator operates on a single proposition.
¬P is read as *not P*.

Its function is to reverse the value of the proposition.
If P is true, then ¬P is false.
If P is false then ¬P is true.

For Example:
If P represents the statement *it is raining*.
then ¬P represents the statement *it is not raining*.

The truth table for the NOT operator is shown in Fig. 5.3.

This is known as **negation.**

Note: you might also come across the notation ~P and !P to mean NOT P.

Fig. 5.3 The truth table for NOT

P	¬P
T	F
F	T

5.4 Logical Operators (Connectives)

Worked Example 3

Let P be "It is cold" and Q be "It is raining". Give simple sentences which represent the following statements:

(a) $\neg P$
(b) $P \wedge Q$
(c) $P \vee Q$
(d) $Q \vee \neg P$
(e) $\neg P \wedge \neg Q$
(f) $\neg \neg Q$

Solution

(a) It is not cold.
(b) It is cold and it is raining.
(c) It is cold or it is raining.
(d) It is raining or it is not cold.
(e) It is not cold and it is not raining.
(f) It is raining.

Worked Example 4

Let P be "She is tall" and Q be "She is intelligent". Express each of the following statements symbolically:

(a) She is tall and intelligent.
(b) She is tall, but not intelligent.
(c) It is false that she is tall or intelligent.
(d) She is neither tall nor intelligent.
(e) She is tall, or she is short and intelligent.
(f) It is not true that she is short or unintelligent.

Solution

(a) $P \wedge Q$
(b) $P \wedge \neg Q$
(c) $\neg(P \vee Q)$
(d) $\neg P \wedge \neg Q$
(e) $P \vee (\neg P \wedge Q)$
(f) $\neg (\neg P \vee \neg Q)$

1. Enter all the possible combinations in columns 1 to 3.

2. Using the values for *P* and *Q* in columns 1 and 2, complete column 4 by using the rules for AND.

3. Complete column 5 by negating the values for *R* in column 3.

4. Using the values in columns 4 and 5, complete column 6 by using the rules for OR.

1	2	3	4	5	6
P	Q	R	$P \wedge Q$	$\neg R$	$(P \wedge Q) \vee \neg R$
T	T	T	T	F	T
T	T	F	T	T	T
T	F	T	F	F	F
T	F	F	F	T	T
F	T	T	F	F	F
F	T	F	F	T	T
F	F	T	F	F	F
F	F	F	F	T	T

Fig. 5.4 Constructing a truth table

5.5 Constructing Truth Tables

Consider the following expression: $(P \wedge Q) \vee \neg R$.

This expression contains three variables, *P*, *Q* and *R*; the overall value of the expression (true or false) will depend on the value of these three variables.

We can construct a truth table to show all possible combinations. In general if there are *n* variables the number of rows in the table will be 2^n.

In this case there are 3 variables so the number of rows will be 2^3 or 8.

The method for construction truth tables is described in Fig. 5.4.

5.6 Logical Equivalence

Two compound statements are said to be **logically equivalent** if they have identical truth tables. We use the symbol ≡ to indicate that two statements are identical.

Worked Example 6
Prove the following identity: $\neg(P \wedge Q) \equiv \neg P \vee \neg Q$.

Solution
See Fig. 5.5.

The last column of each truth table is identical – so both statements are equivalent.

5.7 De Morgan's Law

In the previous section we showed that:

$$\neg(P \wedge Q) \equiv \neg P \vee \neg Q$$

5.7 De Morgan's Law

The left hand side

P	Q	$P \wedge Q$	$\neg(P \wedge Q)$
T	T	T	F
T	F	F	T
F	T	F	T
F	F	F	T

The right hand side

P	Q	$\neg P$	$\neg Q$	$\neg P \vee \neg Q$
T	T	F	F	F
T	F	F	T	T
F	T	T	F	T
F	F	T	T	T

Fig. 5.5 Solution to worked example 6

It can also be shown that:

$$\neg(P \vee Q) \equiv \neg P \wedge \neg Q$$

Together these laws are known as **De Morgan's Law**, and you will remember a similar law from set theory. As we shall see later, the laws of propositional algebra and the laws of set theory are **isomorphic**—which means that "they have the same shape".

De Morgan's Law makes sense—in a previous worked example, we were asked to express the following statements symbolically, where P represented "She is tall" and Q represented "She is intelligent".

- It is false that she is tall or intelligent.
- She is neither tall nor intelligent.

In fact these statements mean exactly the same thing, and our answers were:

- $\neg(P \vee Q)$
- $\neg P \wedge \neg Q$

And by de Morgan's law these two statements are identical!

Worked Example 7
Use De Morgan's law to show that:

$$\neg(\neg P \vee (P \wedge Q)) \equiv P \wedge (\neg P \vee \neg Q)$$

Solution

$$\neg(\neg P \vee (P \wedge Q))$$
$$\equiv \neg\neg P \wedge \neg(P \wedge Q)$$
$$\equiv P \wedge (\neg P \vee \neg Q)$$

The left hand side			
P	Q	P∨Q	¬(P∨Q)
T	T	T	F
T	F	T	F
F	T	T	F
F	F	F	T

The right hand side				
P	Q	¬P	¬Q	¬P∧¬Q
T	T	F	F	F
T	F	F	T	F
F	T	T	F	F
F	F	T	T	T

Fig. 5.6 Solution to worked example 8

Worked Example 8
Use truth tables to prove the second version of De Morgan's law:

$$\neg(P \vee Q) \equiv \neg P \wedge \neg Q$$

Solution
See Fig. 5.6.
The last columns in both truth tables are the same, thus proving the identity.

5.8 Commutativity and Associativity

Both the AND operator and the OR operator are commutative:

$$P \wedge Q \equiv Q \wedge P$$
$$P \vee Q \equiv Q \vee P$$

Both these operators are also associative:

$$(P \wedge Q) \wedge R \equiv P \wedge (Q \wedge R)$$
$$(P \vee Q) \vee R \equiv P \vee (Q \vee R)$$

You can prove the above identities by constructing truth tables.

5.9 The Implication Operator (⇒)

If P and Q are propositions then:

$$P \Rightarrow Q \text{ means : IF } P \text{ THEN } Q.$$
$$\text{or alternatively : } P \text{ IMPLIES } Q.$$

Fig. 5.7 The truth table for IMPLIES

P	Q	P ⇒ Q
T	T	T
T	F	F
F	T	T
F	F	T

The truth table for IMPLIES is given in Fig. 5.7.

If P is the statement "It is sunny", and Q is the statement "I go shopping", then $P \Rightarrow Q$ is the statement "If it is sunny, I go shopping".

Look carefully at the truth table:

Line 1: If the compound statement is true, then if it is sunny, I definitely go shopping.

Line 2: If it is sunny and I do not go shopping, the compound statement must be false.

Lines 3 and 4: If the compound statement is true, but it is not sunny, then maybe I go shopping (line 3) or maybe I do not (line 4).

The statement $P \Rightarrow Q$ being true tells us nothing about what happens if P is false (in this case it is not sunny).

But if P is true then it *guarantees* that Q is true.

P is a *sufficient* condition for Q, but P is not a *necessary* condition for Q.

If $P \Rightarrow Q$ is true, then P is true only if Q is true - if I do not go shopping, it isn't sunny.

For this reason we can interpret $P \Rightarrow Q$ as: P **only if** Q.

Q is a *necessary* condition for P.

5.10 The Equivalence Operator (⇔)

If P and Q are propositions then:

$P \Leftrightarrow Q$ means: P IS EQUIVALENT TO Q.

The truth table for ⇔ is presented in Fig. 5.8.

If P is the statement "It is sunny", and Q is the statement "I go shopping", then $P \Leftrightarrow Q$ is the statement "If it is sunny, I go shopping, and if I go shopping, it is sunny".

Fig. 5.8 The truth table for IS EQUIVALENT TO

P	Q	P ⇔ Q
T	T	T
T	F	F
F	T	F
F	F	T

Fig. 5.9 Solution to worked example 9

P	Q	$\neg P$	$\neg Q$	$\neg P \vee \neg Q$	$Q \Leftrightarrow (\neg P \vee \neg Q)$
T	T	F	F	F	F
T	F	F	T	T	F
F	T	T	F	T	T
F	F	T	T	T	F

Look carefully at the truth table. Line 3 now gives a result of false. So if the statement is true (lines 1 and 4), then if it is sunny I go shopping, if it is not sunny I do not go shopping.

The equivalence operator takes away the uncertainty of the implies operator. I go shopping only if it is sunny; also it is sunny only if I go shopping.

For this reason we can interpret $P \Leftrightarrow Q$ as:
P **if and only if** Q (sometimes written as P iff Q).
P is a *necessary and sufficient* condition for Q.
Q is a *necessary and sufficient* condition for P.

Worked Example 9
Construct a truth tables for the following expression:

$$Q \Leftrightarrow (\neg P \vee \neg Q)$$

Solution
See Fig. 5.9.

5.11 The Equivalence Operator Versus Logical Equivalence

It is important not to confuse the equivalence operator ($\Leftrightarrow$) with logical equivalence ($\equiv$).

The first is an *operator* that joins two propositions: each of the two propositions can be true or false, and the truth of the compound statement is determined from the truth table.

The second is used with two statements (usually compound statements) and means that they are identical because they have the same truth table.

5.12 Order of Precedence of Logical Operators

Just as with arithmetic operators, it is not always immediately clear in which order to do things. The order in which operations should be performed is as follows:

$$\neg$$
$$\wedge$$
$$\vee$$
$$\Rightarrow$$
$$\Leftrightarrow$$

However, it is recommended that brackets are always used in any compound statement that contains several operators.

For example, if you were asked to construct a truth table for this expression:

$$P \Rightarrow Q \vee R \wedge Q$$

you would start with a column for $R \wedge Q$, then do $Q \vee R \wedge Q$, and finally $P \Rightarrow Q \vee R \wedge Q$. However, you will agree that it is a lot clearer if you write it like this:

$$P \Rightarrow (Q \vee (R \wedge Q))$$

5.13 Tautologies and Contradictions

A **tautology** is an expression for which all rows of the truth table evaluate to true.
For example: $P \vee \neg P$.
The truth table for this expression appears in Fig. 5.10.
A **contradiction** is an expression for which all rows of the truth table evaluate to false.
For example: $P \wedge \neg P$.
The truth table for this expression is shown in Fig. 5.11.

Fig. 5.10 An example of a tautology

P	$\neg P$	$P \vee \neg P$
T	F	T
F	T	T

Fig. 5.11 An example of a contradiction

P	$\neg P$	$P \wedge \neg P$
T	F	F
F	T	F

P	Q	$P \Rightarrow Q$	$P \wedge (P \Rightarrow Q)$	$(P \wedge (P \Rightarrow Q)) \Rightarrow Q$
T	T	T	T	T
T	F	F	F	T
F	T	T	F	T
F	F	T	F	T

Fig. 5.12 Solution to worked example 10

Fig. 5.13 The truth table for EXCLUSIVE OR

P	Q	$P \oplus Q$
T	T	F
T	F	T
F	T	T
F	F	F

Worked Example 10
By constructing a truth table, determine whether the following expression is a tautology, a contradiction or neither:

$$(P \wedge (P \Rightarrow Q)) \Rightarrow Q$$

Solution
See Fig. 5.12.
 All the rows in the final column are true, so the expression is a tautology.

5.14 The EXCLUSIVE OR Operator ($\oplus$)

With the OR operator, all that is required for the compound statement to be true is that either one of the simple statements is true.

With the EXCLUSIVE OR operator (sometimes known as XOR), the compound statement is true only when one (and only one) proposition is true. If both are true, then the compound statement is false. The truth table is shown in Fig. 5.13.

You might observe that XOR is actually closer to the way we use the word "or" in natural language than is OR.

If you said "Tonight we are going to the cinema or we are going ice-skating", most people would assume you were doing one or the other but not both!

5.15 Converse, Inverse and Contrapositive

- $Q \Rightarrow P$ is referred to as the **converse** of $P \Rightarrow Q$

5.16 Algebra of Propositions

P	Q	P⇒Q
T	T	T
T	F	F
F	T	T
F	F	T

P	Q	¬P	¬Q	¬Q ⇒ ¬P
T	T	F	F	T
T	F	F	T	F
F	T	T	F	T
F	F	T	T	T

Fig. 5.14 The contrapositive is equivalent to the original statement

- $\neg P \Rightarrow \neg Q$ is referred to as the **inverse** of $P \Rightarrow Q$
- $\neg Q \Rightarrow \neg P$ is referred to as the **contrapositive** of $P \Rightarrow Q$

The contrapositive is equivalent to the original statement, as can be seen from the truth tables in Fig. 5.14.

5.16 Algebra of Propositions

You will see with the laws that follow that there are parallels with set theory. As we noted earlier, there is an **isomorphism** between set theory and propositional logic.

We have already seen that AND and OR are commutative and associative, as are intersection and union in set theory.

The following laws will also look familiar:

Idempotent Laws

$$P \wedge P \equiv P$$
$$P \vee P \equiv P$$

Identity Laws

$$P \wedge F \equiv F$$
$$P \wedge T \equiv P$$
$$P \vee F \equiv P$$
$$P \wedge T \equiv T$$

Complement Laws

$$P \vee \neg P \equiv T$$
$$P \wedge \neg P \equiv F$$
$$\neg T \equiv F$$
$$\neg F \equiv T$$

Distributive Law

$$P \vee (Q \wedge R) \equiv (P \vee Q) \wedge (P \vee R)$$
$$P \wedge (Q \vee R) \equiv (P \wedge Q) \vee (P \wedge R)$$

De Morgan's Law

$$\neg(P \vee Q) \equiv \neg P \wedge \neg Q$$
$$\neg(P \wedge Q) \equiv \neg P \vee \neg Q$$

You can see that $\wedge$ and $\vee$ correspond to $\cap$ and $\cup$ in set theory, while T and F correspond to U and $\emptyset$. In addition, negation in logic corresponds to complement in set theory.

Worked Example 11
Use the distributive law to simplify the following expression:

$$P \vee (\neg P \wedge Q)$$

Solution

$P \vee (\neg P \wedge Q)$
$\equiv (P \vee \neg P) \wedge (P \vee Q)$ (Distributive Law)
$\equiv T \wedge (P \vee Q)$ (Complement Law)
$\equiv P \vee Q$ (Identity Law)

5.17 Some More Algebra

Look at this expression: $\neg P \vee Q$.
The truth table for this expression appears in Fig. 5.15.
This should ring some bells - it is also the truth table for $P \Rightarrow Q$.
So we have the following identity: $P \Rightarrow Q \equiv \neg P \vee Q$ **Identity 1**
Now let's see what happens if we negate $P \Rightarrow Q$:
Using identity 1: $\neg(P \Rightarrow Q) \equiv \neg(\neg P \vee Q)$

Fig. 5.15 The truth table for $\neg P \vee Q$

P	Q	¬P	¬P∨Q
T	T	F	T
T	F	F	F
F	T	T	T
F	F	T	T

We can apply De Morgan's law to the right-hand side of the identity:

$$\neg(\neg P \vee Q) \equiv P \wedge \neg Q$$

So $\neg(P \Rightarrow Q) \equiv P \wedge \neg Q$ **Identity 2**

Worked Example 12
Use algebra to show that: $(P \wedge Q) \Rightarrow \neg Q \equiv \neg P \vee \neg Q$.

Solution

$(P \wedge Q) \Rightarrow \neg Q$
$\equiv \neg(P \wedge Q) \vee \neg Q$ Using identity 1
$\equiv \neg P \vee \neg Q \vee \neg Q$ De Morgan's law
$\equiv \neg P \vee (\neg Q \vee \neg Q)$ Using the fact that $\vee$ is associative
$\equiv \neg P \vee \neg Q$ Idempotent law

Worked Example 13
Negate the following statement and simplify the answer.

$$(P \wedge Q) \Rightarrow R$$

Solution

$\neg((P \wedge Q) \Rightarrow R)$
$\equiv (P \wedge Q) \wedge \neg R$ Using identity 2
$\equiv P \wedge Q \wedge \neg R$ Using the fact that $\wedge$ is associative

5.18 Application to Computing

5.18.1 Programming

One of the most common statements used in program code is the *conditional* statement. An example in Python might look like this:

```
if x < 10 or y > 100:
  # statements go here
```

In Java the above code fragment would look like this:

```
if(x < 10 || y > 100)
{
 // statements go here
}
```

The || symbol in Java (and many other languages) means OR, corresponding to our logical OR. The symbol && is used in Java for the logical AND.

The compiler will use the rules we have learnt to evaluate the statement that follows **if** and thereby determine if it is true or false.

If it is true, the statements that are indented (in the Python example) or contained in the braces (in the Java example) will be executed – otherwise, nothing will happen and the program will move on to the next bit of code following the indentation or braces. In this way, the **if** statement implements the implication operator, $\Rightarrow$. It does not say anything about what to do if the statement is false (other than do nothing and move on).

5.18.2 Digital Electronics and Logic Gates

Modern electronic devices such as mobile telephones and computers depend on digital electronics. At the heart of digital electronics are **logic circuits** - such circuits depend on pulses of electricity to make the circuit function. For example, if the current is high, this is represented as '1', if the current low, this is represented as '0'. This makes it possible to represent and process binary numbers in a computer system.

Mathematical logic is based on a two-valued system (true and false). In digital electronics we also have a two-valued system but have 1 and 0 instead of true and false. Many of the operations that need to be performed on binary numbers within a computer system mirror the logical operations such as NOT, AND, OR. We are able to build **logic gates** that perform these, and other more complex, operations. Three examples appear in Figs. 5.16, 5.17 and 5.18.

NOT gate

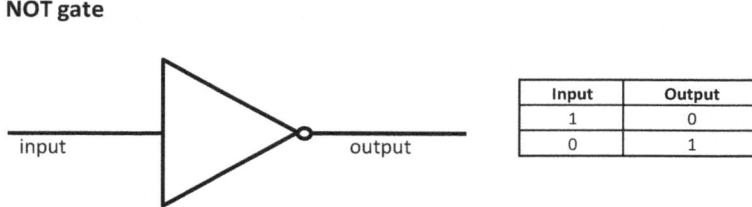

Fig. 5.16 The NOT Gate

OR gate

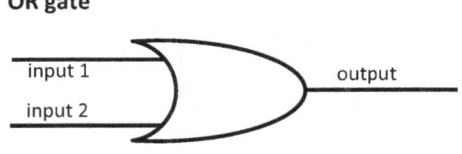
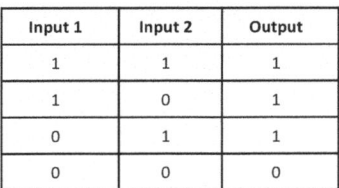

Fig. 5.17 The OR gate

AND gate

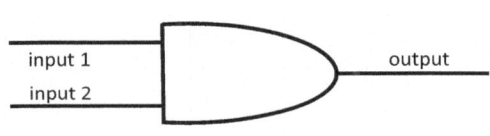

Input 1	Input 2	Output
1	1	1
1	0	0
0	1	0
0	0	0

Fig. 5.18 The AND gate

5.19 Three-Valued Logic

Classical logic assumes that all expressions evaluate to true or false. However, in reality, this is not always the case when evaluating an expression, because sometimes an expression can be undefined. For example, the expression 0/0. Also, undefined terms are very common in programming situations—for example, when a variable is first declared and has not yet been assigned a value.

A system of three-valued logic has therefore been developed to deal with the possibility that a proposition could have the value **true**, **false** or **undefined**.

The truth tables for AND and OR in this system are shown Figs. 5.19 and 5.20.

Looking at the truth table for AND, we see that if either P or Q is false, then the compound statement is false, even if the other proposition is undefined - because

Fig. 5.19 Three-valued logic: truth table for AND

P	Q	$P \wedge Q$
T	T	T
T	F	F
T	UNDEFINED	UNDEFINED
F	T	F
F	F	F
F	UNDEFINED	F
UNDEFINED	T	UNDEFINED
UNDEFINED	F	F
UNDEFINED	UNDEFINED	UNDEFINED

Fig. 5.20 Three-valued logic: truth table for OR

P	Q	P∨Q
T	T	T
T	F	T
T	UNDEFINED	T
F	T	T
F	F	F
F	UNDEFINED	UNDEFINED
UNDEFINED	T	T
UNDEFINED	F	UNDEFINED
UNDEFINED	UNDEFINED	UNDEFINED

both have to be true in order for the compound statement to be true. However, if either P or Q is undefined and the other one is true, then the compound statement is undefined, because its overall value will depend on the value of the undefined proposition.

Similarly, in the table for OR, if just one of the propositions, P or Q, is true, then the compound statement is true - the other proposition being undefined doesn't matter in this case. However, if one proposition is false and the other undefined then the compound statement is undefined because it relies on the value of the undefined proposition.

5.20 Notation in other Texts

Unfortunately, you will see different notation used in different texts. Some texts use single-barred arrows for implication and equivalence:

So instead of $\Rightarrow$ and $\Leftrightarrow$

you will see $\rightarrow$ and $\longleftrightarrow$.

To make matters worse, in texts that use this notation you will often see $\Leftrightarrow$ used for logical equivalence instead of $\equiv$.

In fact, some mathematicians actually make a distinction between the single-barred arrow and the double one. They argue that you should use the single-barred arrows for the logical connectives that join propositions - but you should use the double-barred arrow for an **assertion** – that is a statement that is *true*.

In this notation:

$P \rightarrow Q$ means that P implies Q, and P and Q can take any values.

But: $P \Rightarrow Q$ means that P implies Q is *true* (so in this case we can't have P true and Q false).

5.21 Exercises

1. If P, Q and R represent the following statements:
    P: $2 \times 25 = 50$.
    Q: Germany is in Asia.

5.21 Exercises

    R: 3.65 is an integer
    What is the value of: (a) $P \wedge R$ (b) $P \vee Q$ (c) $Q \vee R?$

2. Let P be "It is summer" and Q be "Leon is playing tennis". Give simple sentences which represent the following statements:
   (a) $\neg P$
   (b) $P \wedge \neg Q$
   (c) $\neg P \vee Q$
   (d) $\neg \neg Q$

3. Let P be "She is a scientist" and Q be "She is intelligent". Express each of the following statements symbolically:
   (a) She is intelligent, but she is not a scientist.
   (b) She is a scientist, and she is intelligent.
   (c) She is a scientist, or she is not intelligent.
   (d) It is not true that she is a scientist or that she is intelligent.

4. Construct a truth table for the following expression:

$$\neg(P \wedge Q) \vee \neg Q$$

5. Show that the expression $P \Rightarrow (P \vee Q)$ is a tautology by constructing a truth table.

6. Consider the following statement: $P \Rightarrow \neg Q$
   What is
   (a) the converse;
   (b) the inverse;
   (c) the contrapositive?

7. Use De Morgan's law to show that:

$$\neg(\neg P \wedge (P \vee Q)) \equiv P \vee (\neg P \wedge \neg Q)$$

8. Use the distributive law to simplify the following expression:

$$\neg Q \wedge (\neg P \vee Q)$$

9. Negate the following expression, and simplify your answer (Hint: Use De Morgan's Law):

$$(P \Rightarrow Q) \wedge Q$$

10. In question 5 you drew a truth table to show that $P \Rightarrow (P \vee Q)$ is a tautology. Now do this using algebra.

11. Draw a truth table for the following expression, using **3-valued** logic:

$$P \vee \neg Q$$

Predicate Logic and Proofs

At the end of this chapter you should be able to:

- define the term **predicate**;
- explain the notion of a predicate as a Boolean function;
- give values to the variables in predicates by **substitution** and **quantification**;
- negate quantified predicates;
- state the following laws of natural deduction: **modus ponens, modus tollens, chain rule, AND-elimination, AND-introduction, OR-introduction, universal instantiation, existential generalization**;
- use the above laws to perform proofs by **natural deduction**;
- perform simple proofs by **mathematical induction**;
- explain how mathematical logic is used in formal methods of software engineering.

6.1 Introduction

Having completed our work on propositions, we are now in a position to move on to the other branch of logic, namely predicate logic. We will discover that a predicate is a function that contains variables, and outputs a value of TRUE or FALSE. We will learn the notation used for predicates, we will find out how to give values to the variables that a predicate contains, and we will then move on to study some laws which enable us to perform mathematical proofs.

6.2 Predicate Logic

We have learnt that a *proposition* is an expression that has a value of TRUE or FALSE. A **predicate** is also an expression that has a value of TRUE or FALSE. However, a predicate contains *variables*; we don't know if the predicate is TRUE or FALSE until we know the value of the variables.

6.3 Examples of Predicates

$P(x)$: x is a prime number (usually pronounced P of x)
$T(x, y)$: x is taller than y
$D(x)$: x is a duck

As you can see, we don't know whether $P(x)$, for example, is TRUE or FALSE until we know the value of x.

For example: $P(5)$ is TRUE
but: $P(10)$ is FALSE

We should note that predicates are *functions*. A predicate, just like a function, accepts variables and produces an output. In this case the output is a truth value—either TRUE or FALSE.

So a predicate is *a truth-valued function*.

If A is the set of people, then the predicate $T(x, y)$ above would have the following signature:

$$T: A \times A \to \mathbb{B}$$

$\mathbb{B}$ is a special, pre-defined set known as the **Boolean** set:

$$\mathbb{B} = \{TRUE, FALSE\}.$$

6.4 The Domain of Discourse

When we have a predicate such as $P(x)$, it is always important to know the set of values from which is x drawn. This set is known as the **domain of discourse**. For a predicate like "x is an even number", our domain of discourse might be natural numbers. For a predicate such as "studies physics", our domain of discourse could be people or students.

We can make our domain of discourse clear in one of two ways:

- we can simply state it in advance;
- we can incorporate it into the definition, as we shall see below.

6.5 Giving Values to the Variables

Once we give a value to the variables, we know whether the predicate is TRUE or FALSE. This is also known as **binding** the variables. There are two ways of doing this:

6.5.1 Substitution

As we saw previously, we can simply substitute specific values for the variables. Using the examples from Sect. 6.3:

$P(7)$:	7 is a prime number
T(Mary, Ahmed):	Mary is taller than Ahmed
D(Basil):	Basil is a duck.

6.5.2 Quantification

A **quantifier** is a device for making a statement about a *set* of values, not just one value.

There are three quantifiers that we can use, each with its own symbol.

6.5.2.1 The Universal Quantifier, ∀
This quantifier enables us to make a statement about *all* the elements in a particular set.

If $P(x)$ is a predicate:

$$\forall x \cdot P(x) \text{ reads: For all } x, P(x) \text{ is true.}$$

This assumes we have already stated our domain of discourse in advance.

If we had not, we would have to include it in the statement – for example if the domain of discourse was the set A, we would write:

$$\forall x \in A \cdot P(x)$$

Example
If $M(x)$ is the predicate *x chases mice* (defined over the set of cats), then
$\forall x \cdot M(x)$ reads: All cats chase mice.

6.5.2.2 The Existential Quantifier ∃
This quantifier makes a statement about *at least one* of the elements in a particular set.

If $P(x)$ is a predicate:

$\exists x \cdot P(x)$ reads: There exists at least one x, for which $P(x)$ is true.

As before, this assumes we have already stated our domain of discourse in advance—otherwise (if our domain of discourse was the set A) we would write:

$$\exists x \in A \cdot P(x)$$

Example
If $M(x)$ is the predicate *x chases mice* (defined over the set of cats), then

$\exists x \cdot M(x)$ reads: There is at least one cat that chases mice.

6.5.2.3 The Unique Existential Quantifier ∃!
This quantifier makes a statement about *one and only one* of the elements in a particular set:
If $P(x)$ is a predicate:

$\exists! x \cdot P(x)$ reads: There exists one and only one x, for which $P(x)$ is true.

Example
If $M(x)$ is the predicate *x chases mice*, (defined over the set of cats), then

$\exists! x \cdot M(x)$ reads: There is one and only one cat who chases mice.

Worked Example 1

(a) $B(x)$ is the predicate *x can bark*, defined over the domain of dogs.
   Write the following statements in words:
   (i) $B(\text{Rover})$
   (ii) $\forall x \cdot B(x)$
   (iii) $\exists x \cdot B(x)$
   (iv) $\exists! x \cdot B(x)$
   (v) $\neg B(\text{Rover})$
   (vi) $\neg \forall x \cdot B(x)$
   (vii) $\neg \exists x \cdot B(x)$
(b) If we had not stated the domain of discourse in advance, how would we have written part (ii), using D to represent the set of dogs?

6.5 Giving Values to the Variables 89

Solution

(a)

  (i) Rover can bark.
  (ii) All dogs can bark.
  (iii) There is at least one dog that can bark.
  (iv) There is one and only one dog that can bark.
  (v) Rover cannot bark.
  (vi) It is not true that all dogs can bark.
  (vii) There does not exist a dog that can bark.

(b) $\forall x \in D \cdot B(x)$

Worked Example 2

$B(x)$ is the predicate *x can bark*, and $T(x)$ is the predicate *x has a tail*, both defined over the domain of dogs.

Write the following in words:

(a) $\forall x \cdot (B(x) \wedge T(x))$
(b) $\exists x \cdot (B(x) \wedge \neg T(x))$
(c) $\forall x \cdot B(x) \wedge \exists x \cdot \neg T(x)$

Solution

(a) All dogs can bark and have tails.
(b) There is at least one dog that can bark and that has no tail.
(c) All dogs can bark, and there is at least one dog that has no tail.

Worked Example 3

$D(x)$ is the predicate *x is a duck*, and $S(x)$ is the predicate *x can swim*, both defined over the domain of animals.

(a) Write the following statements in words:
  (i) $\forall x \cdot (D(x) \Rightarrow S(x))$
  (ii) $\exists x \cdot (D(x) \wedge \neg S(x))$
  (iii) $D(\text{Basil}) \wedge \exists! x \cdot S(x)$
(b) Write the following statements in symbols:
  (i) If Basil is a duck, then all animals can swim.
  (ii) Only ducks can swim (every animal is a duck or it cannot swim).
  (iii) There is one and only one duck that can swim.

Solution

(a)
    (i) All ducks can swim (if x is a duck then x can swim).
    (ii) There exists an animal that is a duck and cannot swim (there is at least one duck that cannot swim).
    (iii) Basil is a duck and there is one and only one animal that can swim.

(b)
    (i) $D(\text{Basil}) \Rightarrow \forall x \cdot S(x)$
    (ii) $\forall x \cdot (D(x) \vee \neg S(x))$
    (iii) $\exists ! x \cdot (D(x) \wedge S(x))$

6.5.3 Negating Quantified Predicates

Consider the following predicate, defined over the domain of students at a particular university:

$M(x)$: x is good at maths

The following statement means that it is not true that all students are good at maths:

$$\neg \forall x \cdot M(x)$$

This means that there must be at least one student that is not good at maths:

$$\exists x \cdot \neg M(x)$$

Therefore: $\neg \forall x \cdot M(x) \equiv \exists x \cdot \neg M(x)$

Similarly, the following statement means that there does not exist a single student who is good at maths:

$$\neg \exists x \cdot M(x)$$

This is the same as saying that all students are poor at maths:

$$\forall x \cdot \neg M(x)$$

Therefore: $\neg \exists x \cdot M(x) \equiv \forall x \cdot \neg M(x)$

Worked Example 4

Negate the predicate $\forall x \cdot (\neg P(x))$.

Solution

$$\neg \forall x \cdot (\neg P(x)) \equiv \exists x \cdot \neg(\neg P(x))$$
$$\equiv \exists x \cdot P(x)$$

Fig. 6.1 The truth table for IMPLIES

P	Q	$P \Rightarrow Q$
T	T	T
T	F	F
F	T	T
F	F	T

6.6 Proof by Natural Deduction

Mathematical logic gives us the ability to make **logical arguments**. The way we do this is to make a number of premises, and to show that if these **premises** are correct, a certain **conclusion** follows.

The notation for this is to place the premises above a horizontal line, and the conclusion below it.

For example:

$$\frac{P; Q; R}{S}$$

This means that if P, Q and R are true, then it follows that S is also true.

To help us with this, there are some basic laws that follow from the definitions we learnt in the previous chapter.

6.6.1 Modus Ponens

Imagine these two statements are true:
If it's Wednesday I have eggs for breakfast.
It is Wednesday.
It follows that I have eggs for breakfast.
This is the law of **modus ponens**. It is stated formally like this:

$$\frac{P \Rightarrow Q; P}{Q}$$

It follows directly from our understanding of the implication operator and its truth table (reproduced in Fig. 6.1).

It tells us that if $P \Rightarrow Q$ is true and P is also true, then Q must be true (the first line of the truth table).

6.6.2 Modus Tollens

Imagine these two statements are true:
If it's Wednesday I have eggs for breakfast.
I do not have eggs for breakfast.

It follows that it is not Wednesday.
This is the law of **modus tollens**. It is stated formally like this:

$$\frac{P \Rightarrow Q;\ \neg Q}{\neg P}$$

Again, it follows directly from our understanding of the implication operator and its truth table.

It tells us that if $P \Rightarrow Q$ is true and Q is false, then P must be false (the fourth line of the truth table in Fig. 6.1).

6.6.3 The Chain Rule

Imagine these two statements are true:
   If it's Friday I go shopping.
   If I go shopping I wear my yellow hat.
   It follows that if it's Friday I wear my yellow hat.
This is the **chain rule**. It is stated formally like this:

$$\frac{P \Rightarrow Q;\ Q \Rightarrow R}{P \Rightarrow R}$$

It arises naturally from our definition of implication.

Worked Example 5
Show that if the following statements are true:

   If the sun is hot then the moon is made of cheese.
   If the moon is made of cheese then grass is blue.
   Grass is not blue.

Then it follows that:

   The sun is not hot.

Solution
First assign letters to the propositions:

S: The sun is hot
M: The moon is made of cheese
G: Grass is blue

6.6 Proof by Natural Deduction

We have to show that:

$$\frac{S \Rightarrow M\,;\,M \Rightarrow G\,;\,\neg G}{\neg S}$$

We do this as follows:

1. $S \Rightarrow M$ Premise
2. $M \Rightarrow G$ Premise
3. $\neg G$ Premise
4. $S \Rightarrow G$ Chain rule on 1 and 2
5. $\neg S$ Modus Tollens on 3 and 4

We have shown that our premises lead to the conclusion $\neg S$: the sun is not hot.

6.6.4 Alternative Names

You might come across the following alternative names for the three rules we have just learnt:
  Modus ponens: the rule of detachment
  Modus tollens: the rule of contraposition
  The chain rule: the rule of syllogism

6.6.5 Some Other Laws

6.6.5.1 The Rule of AND-Elimination
If $P \wedge Q$ is true then it follows from the truth table for AND that P is true and Q is true. This is stated formally as follows:

$$\frac{P \wedge Q}{P} \quad \frac{P \wedge Q}{Q}$$

6.6.5.2 The Rule of AND-Introduction
If P is true and Q is true, it follows from the truth table for AND that $P \wedge Q$ is true. Formally:

$$\frac{P\,;\,Q}{P \wedge Q}$$

6.6.5.3 The Rule of OR-Introduction
If P is true it follows it follows from the truth table for OR that $P \vee Q$ is true, giving us the following law:

$$\frac{P}{P \vee Q}$$

6.6.5.4 Universal Instantiation
If we know something is true for all members of a group, we can conclude it is also true for any arbitrary member of this group:

$$\frac{\forall x \cdot P(x)}{P(c)}$$

for any c in the domain.

6.6.5.5 Existential Generalization
If we know that $P(c)$ is true for some constant c, then there exists an element for which P is true. Thus, we can conclude $\exists x \cdot P(x)$:

$$\frac{P(c)}{\exists x \cdot P(x)}$$

Worked Example 6
Show that if the following statements are true:

If maths is easy and physics is hard then apples are green;
If physics is hard then maths is easy;
Physics is hard.

then it follows that:

Apples are green.

Solution
We can define the following propositions:

P: Maths is easy
Q: Physics is hard
R: Apples are green

6.6 Proof by Natural Deduction

We need to show that:

$$\frac{(P \wedge Q) \Rightarrow R;\ Q \Rightarrow P;\ Q}{R}$$

1.	$(P \wedge Q) \Rightarrow R$	Premise
2.	$Q \Rightarrow P$	Premise
3.	Q	Premise
4.	P	Modus ponens on 2, 3
5.	$P \wedge Q$	AND-introduction on 3, 4
6.	R	Modus ponens on 5, 1

We have shown that our premises lead to the conclusion R: apples are green.

Worked Example 7

Show that if the following statements are true:

All dogs can bark;
Rover is a dog;

then it follows that:

Rover can bark.

Solution

We will define the following predicates:

$P(x)$: x is a dog
$Q(x)$: x can bark

and the following constant

r: Rover.

We have to show that:

$$\frac{\forall x \cdot (P(x) \Rightarrow Q(x))}{Q(r)}$$

> 1. $\forall x \cdot (P(x) \Rightarrow Q(x))$ Premise
> 2. $P(r)$ Premise
> 3. $P(r) \Rightarrow Q(r)$ Universal instantiation on 1
> 4. $Q(r)$ Modus ponens on 2, 3
>
> We have shown that our premises lead to the conclusion $Q(r)$: Rover can bark.

6.7 Application to Computing

Logic has a very important role to play in many areas of computer science. These include such areas as artificial intelligence, computer architecture (think back to the logic gates described in Sect. 5.18.2), the development of programming languages, databases, and derivation of algorithms, and the theory of computation.

One particular area is in the development of commercial applications that require high integrity software—that is to say, software that requires a very high level of confidence in its correctness. Such situations include safety critical systems and secure financial systems. **Formal methods** are often used to produce such software. With such methods the specification is written not just in a natural language such as English, but in the language of mathematics. In this way the specification can be far more precise and unambiguous. The mathematics involves the propositional and predicate logic that we have studied here. Using proofs such as those we have studied, the process can be taken further and used to develop the software itself, and help ensure its correctness. The applications developed in this way will be of a higher integrity than those developed by non-formal methods.

Well known formal methods include:

- VDM (Vienna Development Method);
- Z;
- B-Method.

6.8 Proof by Induction

Mathematical induction is a technique that can be used to prove certain statements about natural numbers. It does not use mathematical logic, but it is included here as another example of mathematical proof.

In its simplest form, mathematical induction shows that a statement involving a natural number n holds for all values of n.

The proof involves two steps:

1. The **base step**

6.8 Proof by Induction

Prove that the statement holds for the first natural number n, usually, $n = 0$ or $n = 1$.

2. The **inductive step**

Show that if the statement is true for some natural number k, then it is also true for $k + 1$.

This shows that if the statement is true when $n = 0$ (for example), it is also true when $n = 1$, $n = 2$, $n = 3$ and so on up to infinity.

Worked Example 8
Prove that the following statement holds for all natural numbers:

$$1 + 2 + 3 + \ldots + n = \frac{n(n+1)}{2}$$

Solution

Base Step
Show that it holds when $n = 1$. This is easily shown:

$$1 = \frac{1(1+1)}{2}$$

Inductive Step
Assume the statement is true for some value $n = k$:

$$1 + 2 + 3 + \ldots + k = \frac{k(k+1)}{2} \tag{6.1}$$

Now take the sequence up to $k + 1$:

$$1 + 2 + 3 + \ldots + k + k + 1$$

Substituting from Eq. 6.1 this becomes:

$$\frac{k(k+1)}{2} + k + 1$$
$$= \frac{k^2 + k + 2k + 2}{2}$$
$$= \frac{k^2 + 3k + 2}{2}$$
$$= \frac{(k+1)(k+2)}{2}$$

This can be written as:
$$\frac{(k+1)((k+1)+1)}{2}$$

Thus if it true for $n = k$, it is true for $n = k + 1$, and since it is true for $n = 1$, it is true for all n.

Worked Example 9
Prove that for all $n \geq 1$, the expression $n^3 + 2n$ is divisible by 3.

Solution

Base Step
Show that it holds when $n = 1$:
$1^3 + 2 \times 1 = 3$, which is divisible by 3.

Inductive Step
Assume the statement is true for some value $n = k$:
$$k^3 + 2k = 3M \qquad (6.2)$$

where M is a positive integer.
When $n = k + 1$, the expression becomes:
$$\begin{aligned}
&(k+1)^3 + 2(k+1) \\
&= (k+1)(k+1)^2 + 2(k+1) \\
&= (k+1)(k^2 + 2k + 1) + 2(k+1) \\
&= (k^3 + 3k^2 + 3k + 1) + 2(k+1) \\
&= (k^3 + 3k^2 + 5k + 3) \\
&= (k^3 + 2k) + (3k + 3k^2 + 3)
\end{aligned}$$

Substituting from Eq. 6.2 this becomes:
$$\begin{aligned}
&3M + (3k + 3k^2 + 3) \\
&= 3(M + k + k^2 + 1)
\end{aligned}$$

This is divisible by 3.
Thus if it true for $n = k$, it is true for $n = k + 1$, and since it is true for $n = 1$, it is true for all $n \geq 1$.

6.9 Exercises

1. $D(x)$ is the predicate *x is a duck*, defined over the domain of animals.
   (a) Write the following statements in words:
      (i) $D(\text{BASIL})$
      (ii) $\forall x \cdot \neg D(x)$
      (iii) $\exists x \cdot D(x)$
      (iv) $\exists! x \cdot \neg D(x)$
      (v) $\neg \forall x \cdot D(x)$
   (b) If we had not stated the domain of discourse in advance, how would we have written part (ii), using A to represent the set of animals?
2. $B(x)$ is the predicate *x is a bird*, and $F(x)$ is the predicate *x can fly*, both defined over the domain of animals.
   (a) Write the following statements in words:
      (i) $\forall x \cdot (B(x) \Rightarrow F(x))$
      (ii) $\exists x \cdot (B(x) \wedge F(x))$
      (iii) $B(\text{JACK}) \vee \exists x \cdot F(x)$
   (b) Write the following statements in symbols:
      (i) If Mary is a bird, then no animal can fly.
      (ii) Only birds can fly (every animal is a bird or it cannot fly).
      (iii) There is one and only one bird that cannot fly.
3. Negate the predicate $\exists x \cdot (\neg P(x) \wedge \neg Q(x))$ by using the universal quantifier ($\forall$) instead of the existential quantifier ($\exists$).
4. Simplify your answer to question 3 by using De Morgan's law.
5. Show that if the following statements are true:

   if I do the ironing I have a cup of tea in the afternoon.
   if I have a cup of tea in the afternoon it is Thursday.
   I do the Ironing.

   Then it follows that:

   it is Thursday.

6. Show that if the following statements are true:

   Bernard is a cat and Susan is a cat;
   Bernard likes watching television.
   If there is at least one cat that likes watching television then the moon is made of cheese.

   then it follows that:

   The moon is made of cheese.

7. Show that if the following statements are true:

   Sam is a snake;
   If Sam can bite then Paris is in France;
   All snakes can bite;

   then it follows that:

   Paris is in France.

8. Prove that the following statement holds for all $n \geq 1$:

   $$2 + 2^2 + 2^3 + 2^4 + \cdots + 2^n = 2^{n+1} - 2$$

Matrices   7

At the end of this chapter you should be able to:

- provide a definition of a **matrix**;
- use appropriate terms to describe different types of matrix (such as **square matrix**, **column vector** and **row vector**);
- **transpose** a matrix;
- add and subtract matrices;
- apply **scalar** and **matrix multiplication**;
- define the **identity matrix** and explain its significance;
- calculate the **determinant** and **inverse** of a 2×2 matrix;
- use the inverse to solve simple matrix equations in the form of $A \times X = B$;
- briefly explain the importance of matrices in computer graphics.

7.1 Introduction

In this chapter, we move on to a completely new topic – the study of **matrices** (singular **matrix**). Matrices are not only used in other branches of mathematics – particularly geometry – but also play an important role in computer science, especially in the areas of computer graphics, cryptography, and data storage.

7.2 Definition and Examples

A matrix is a grid of numbers, consisting of rows and columns. The numbers are enclosed in round or square brackets. Examples of matrices are shown in Fig. 7.1.

Fig. 7.1 Examples of matrices

$$\begin{pmatrix} 3 & 1 & -2 \\ 0 & 1 & 3 \end{pmatrix} \quad \begin{pmatrix} 2 & 4 \\ -1 & 0 \\ 9 & 10 \\ 1 & -3 \end{pmatrix} \quad \begin{pmatrix} 3 & 7 & 0 \\ 20 & 3 & -5 \\ -2 & 1 & 45 \end{pmatrix} \quad \begin{pmatrix} 4 \\ 1 \\ 3 \end{pmatrix} \quad (-11 \ 2 \ 33)$$

A 2 x 3 matrix   A 4 x 2 matrix   A 3 x 3 matrix (An example of a square matrix)   A column matrix (or column vector)   A row matrix (or row vector)

We often use an upper case letter to denote a particular matrix. For example:

$$A = \begin{pmatrix} 1 & 4 \\ 3 & 6 \end{pmatrix}$$

7.3 Matrix Operations

There are a number of operations that can be performed on matrices. We will be studying the following:

- transposition;
- addition/subtraction;
- scalar multiplication;
- matrix multiplication.

7.3.1 Transposition

To transpose a matrix we take each row of the original matrix and make it a column of the transposed matrix.

$$A = \begin{pmatrix} 1 & 2 & 3 \\ 4 & 5 & 6 \end{pmatrix} \quad A^T = \begin{pmatrix} 1 & 4 \\ 2 & 5 \\ 3 & 6 \end{pmatrix}$$

7.3.2 Addition and Subtraction of Matrices

To add (or subtract) two matrices they must be the same shape.
    Then just add (or subtract) the corresponding elements.

7.3 Matrix Operations

Example

$$\begin{pmatrix} 3 & 10 & 1 \\ -2 & 4 & 6 \\ 3 & 1 & 2 \end{pmatrix} + \begin{pmatrix} 5 & 11 & 21 \\ 2 & 5 & 8 \\ 0 & 1 & 12 \end{pmatrix}$$

$$= \begin{pmatrix} 3+5 & 10+11 & 1+21 \\ -2+2 & 4+5 & 6+8 \\ 3+0 & 1+1 & 2+12 \end{pmatrix}$$

$$= \begin{pmatrix} 8 & 21 & 22 \\ 0 & 9 & 14 \\ 3 & 2 & 14 \end{pmatrix}$$

7.3.3 Scalar Multiplication

To multiply a matrix by a scalar quantity (a simple number), just multiply each element by the number.

Example

$$7 \begin{pmatrix} 3 & 10 & 1 \\ -2 & 4 & 6 \\ 3 & 1 & 2 \end{pmatrix}$$

$$= \begin{pmatrix} 3 \times 7 & 10 \times 7 & 1 \times 7 \\ -2 \times 7 & 4 \times 7 & 6 \times 7 \\ 3 \times 7 & 1 \times 7 & 2 \times 7 \end{pmatrix}$$

$$= \begin{pmatrix} 21 & 70 & 7 \\ -14 & 28 & 42 \\ 21 & 7 & 14 \end{pmatrix}$$

7.3.4 Matrix Multiplication

To multiply two matrices together, the number of columns of the first matrix must be equal the number of rows of the second.

Then we multiply the elements of each row of the first matrix by the elements of each column of the second and add them.

This process is explained in Fig. 7.2

Note: Matrix multiplication is not commutative: $A \times B \neq B \times A$.

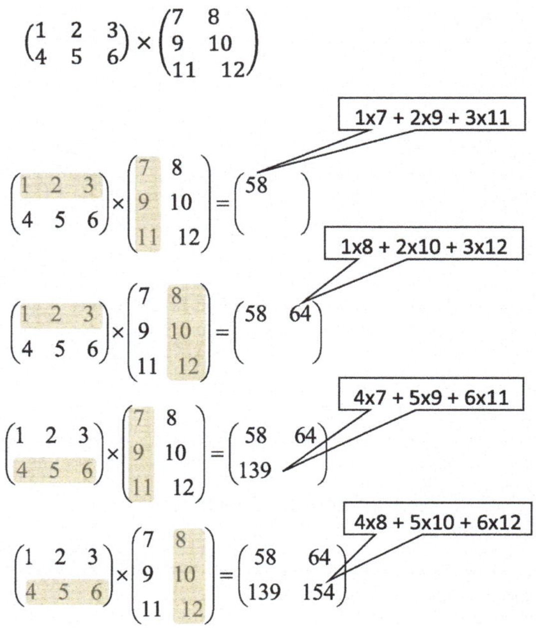

Fig. 7.2 Matrix multiplication

$$\begin{pmatrix} 1 & 2 & 3 \\ 4 & 5 & 6 \end{pmatrix} \times \begin{pmatrix} 7 & 8 \\ 9 & 10 \\ 11 & 12 \end{pmatrix}$$

$$\begin{pmatrix} 1 & 2 & 3 \\ 4 & 5 & 6 \end{pmatrix} \times \begin{pmatrix} 7 & 8 \\ 9 & 10 \\ 11 & 12 \end{pmatrix} = \begin{pmatrix} 58 & \end{pmatrix} \quad \boxed{1 \times 7 + 2 \times 9 + 3 \times 11}$$

$$\begin{pmatrix} 1 & 2 & 3 \\ 4 & 5 & 6 \end{pmatrix} \times \begin{pmatrix} 7 & 8 \\ 9 & 10 \\ 11 & 12 \end{pmatrix} = \begin{pmatrix} 58 & 64 \end{pmatrix} \quad \boxed{1 \times 8 + 2 \times 10 + 3 \times 12}$$

$$\begin{pmatrix} 1 & 2 & 3 \\ 4 & 5 & 6 \end{pmatrix} \times \begin{pmatrix} 7 & 8 \\ 9 & 10 \\ 11 & 12 \end{pmatrix} = \begin{pmatrix} 58 & 64 \\ 139 & \end{pmatrix} \quad \boxed{4 \times 7 + 5 \times 9 + 6 \times 11}$$

$$\begin{pmatrix} 1 & 2 & 3 \\ 4 & 5 & 6 \end{pmatrix} \times \begin{pmatrix} 7 & 8 \\ 9 & 10 \\ 11 & 12 \end{pmatrix} = \begin{pmatrix} 58 & 64 \\ 139 & 154 \end{pmatrix} \quad \boxed{4 \times 8 + 5 \times 10 + 6 \times 12}$$

Worked Example 1

Consider the following matrices:

$$A = \begin{pmatrix} -2 & 2 & 3 \\ 1 & 8 & 4 \end{pmatrix} \quad B = \begin{pmatrix} 3 & 6 & 7 \\ 9 & 8 & 5 \end{pmatrix}$$

Find the value of:

(a) $A + B$ (b) $A - B$ (c) $3A + 2B$ (d) A^T

Solution

(a) $A + B = \begin{pmatrix} -2+3 & 2+6 & 3+7 \\ 1+9 & 8+8 & 4+5 \end{pmatrix} = \begin{pmatrix} 1 & 8 & 10 \\ 10 & 16 & 9 \end{pmatrix}$

(b) $A - B = \begin{pmatrix} -2-3 & 2-6 & 3-7 \\ 1-9 & 8-8 & 4-5 \end{pmatrix} = \begin{pmatrix} -5 & -4 & -4 \\ -8 & 0 & -1 \end{pmatrix}$

(c) $3A + 2B = \begin{pmatrix} -6 & 6 & 9 \\ 3 & 24 & 12 \end{pmatrix} + \begin{pmatrix} 6 & 12 & 14 \\ 18 & 16 & 10 \end{pmatrix} = \begin{pmatrix} 0 & 18 & 23 \\ 21 & 40 & 22 \end{pmatrix}$

(d) $A^T = \begin{pmatrix} -2 & 1 \\ 2 & 8 \\ 3 & 4 \end{pmatrix}$

7.4 The Determinant of a Matrix

Fig. 7.3 Solution to worked example 2

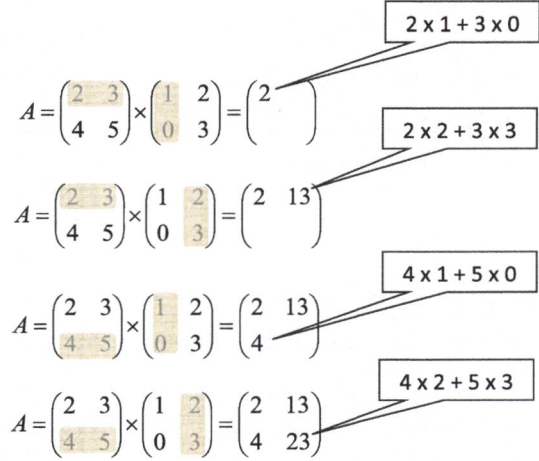

Worked Example 2
Consider the following matrices:

$$A = \begin{pmatrix} 2 & 3 \\ 4 & 5 \end{pmatrix} \quad B = \begin{pmatrix} 1 & 2 \\ 0 & 3 \end{pmatrix}$$

Calculate $A \times B$.

Solution
See Fig. 7.3

It should be noted that the notation for powers used in arithmetic is also used with matrix multiplication. Thus

$$A^2 = A \times A \quad A^3 = A \times A \times A \quad A^4 = A \times A \times A \times A$$

and so on.

7.4 The Determinant of a Matrix

Square matrices have a special property called the **determinant**, which can be very useful in matrix algebra.

The determinant of a matrix A is written as $\det(A)$.

To find the determinant of a 2×2 matrix we multiply each pair of opposite corners and subtract the result (we do the left-to-right diagonal first, and then the right-to-left diagonal). For example:

$$A = \begin{pmatrix} 4 & 3 \\ 2 & 1 \end{pmatrix}$$

$$\det(A) = (4 \times 1) - (3 \times 2)$$
$$= -2$$

Note: You will also see the following notation for the determinant:

$$|A| \quad \begin{vmatrix} 4 & 3 \\ 2 & 1 \end{vmatrix}$$

Worked Example 3
Find the determinant of the following matrix, A:

$$A = \begin{pmatrix} -3 & 3 \\ -1 & 4 \end{pmatrix}$$

Solution

$$\det(A) = (-3 \times 4) - (-1 \times 3)$$
$$= -12 + 3$$
$$= -9$$

7.4.1 Calculating the Determinant of a 3 × 3 Matrix

If:

$$A = \begin{pmatrix} a & b & c \\ d & e & f \\ g & h & i \end{pmatrix}$$

then: $\det(A) = a(ei - fh) - b(di - fg) + c(dh - eg)$

There is a pattern here:

$$\begin{pmatrix} a \\ X \\ \begin{vmatrix} e & f \\ h & i \end{vmatrix} \end{pmatrix} - \begin{pmatrix} b \\ X \\ \begin{vmatrix} d & f \\ g & i \end{vmatrix} \end{pmatrix} + \begin{pmatrix} c \\ X \\ \begin{vmatrix} d & e \\ g & h \end{vmatrix} \end{pmatrix}$$

1. Multiply a by the determinant of the 2×2 matrix that is not in a's row or column.
2. Do the same for b then give it a negative sign;

Fig. 7.4 Identity matrices
$$\begin{pmatrix} 1 & 0 \\ 0 & 1 \end{pmatrix} \qquad \begin{pmatrix} 1 & 0 & 0 \\ 0 & 1 & 0 \\ 0 & 0 & 1 \end{pmatrix} \qquad \begin{pmatrix} 1 & 0 & 0 & 0 \\ 0 & 1 & 0 & 0 \\ 0 & 0 & 1 & 0 \\ 0 & 0 & 0 & 1 \end{pmatrix}$$

3. Do the same for c, but leave it positive.
4. Add them up.

We could also write this as (using the alternative notation for the determinant):

$$|A| = a \times \begin{vmatrix} e & f \\ h & i \end{vmatrix} - b \times \begin{vmatrix} d & f \\ g & i \end{vmatrix} + c \times \begin{vmatrix} d & e \\ g & h \end{vmatrix}$$

Finding the determinant of 3×3 and larger matrices can be very fiddly, but it is possible to use an online calculator. For example: www.matrixcalc.org/en

7.5 Identity Matrices

An **identity matrix** is a square matrix that has 1s along the left-to-right diagonal and 0s everywhere else.

Examples are shown in Fig. 7.4.
In calculations we refer to the identity matrix as I.
When we multiply a matrix by the identity matrix, the original is unchanged:

$$A \times I = A \qquad I \times A = A.$$

(It's like multiplying by 1 in arithmetic).

7.6 The Inverse of a Matrix

The **inverse** of a matrix, A, is a matrix that, when multiplied by A, will give you the identity matrix. The inverse of A is written as A^{-1}.

$$A \times A^{-1} = I$$

Unusually for matrix multiplication, this operation is commutative, so:

$$A^{-1} \times A = I$$

We find that for three matrices, A, B and X:

$$\text{if:} \quad A \times X = B \quad \text{then:} \quad X = A^{-1} \times B$$

Compare this to arithmetic:

$$\text{if: } ax = b \quad \text{then: } x = \frac{1}{a}b.$$

So the inverse acts like the reciprocal in arithmetic.

Why does this work?
Consider the expression: $\quad A \times X = B.$
Multiply both sides by A^{-1}:

$$A^{-1} \times A \times X \doteq A^{-1} \times B.$$
$$\text{but: } \quad A^{-1} \times A = I.$$
$$\text{so: } \quad I \times X = A^{-1} \times B.$$
$$\text{but: } \quad I \times X = X.$$
$$\text{so: } \quad X = A^{-1} \times B.$$

7.6.1 Finding the Inverse

For a 2×2 matrix we calculate the inverse by the following formula:

If $A = \begin{pmatrix} a & b \\ c & d \end{pmatrix}$ Then $A^{-1} = \frac{1}{\det(A)} \begin{pmatrix} d & -b \\ -c & a \end{pmatrix}.$

For higher order matrices it is more complicated, and we will not deal with that in this text.

Example

$$A = \begin{pmatrix} 4 & 1 \\ 2 & 3 \end{pmatrix}$$

$\det(A) = 4 \times 3 - 2 \times 1 = 10$

$$A^{-1} = \frac{1}{10} \begin{pmatrix} 3 & -1 \\ -2 & 4 \end{pmatrix}$$
$$= \begin{pmatrix} 0.3 & -0.1 \\ -0.2 & 0.4 \end{pmatrix}$$

Note:

- Only square matrices have an inverse.
- There is no inverse if the determinant is 0.

7.6 The Inverse of a Matrix

Worked Example 4
Where possible, find the inverse of the following matrices:

a. $A = \begin{pmatrix} 6 \\ 2 \end{pmatrix}$

b. $B = \begin{pmatrix} 3 & 8 \\ 2 & 6 \end{pmatrix}$

c. $C = \begin{pmatrix} -1 & 0 \\ -2 & 0 \end{pmatrix}$

Solution

a. A is not invertible because it is not a square matrix.
b. $\det(B) = 6 \times 3 - 8 \times 2 = 2$.

$$B^{-1} = \frac{1}{2} \begin{pmatrix} 6 & -8 \\ -2 & 3 \end{pmatrix}$$
$$= \begin{pmatrix} 3 & -4 \\ -1 & 3/2 \end{pmatrix}$$

c. $\det(C) = 0$. Therefore C is not invertible.

Worked Example 5
Consider two of the matrices from the previous section:

$$B = \begin{pmatrix} 3 & 8 \\ 2 & 6 \end{pmatrix} \quad A = \begin{pmatrix} 6 \\ 2 \end{pmatrix}$$

If $B \times X = A$, find the value of X.

Solution

$$B \times X = A.$$

Therefore $X = B^{-1} \times A$
In the last question we found that: $B^{-1} = \begin{pmatrix} 3 & -4 \\ -1 & 3/2 \end{pmatrix}$.
Therefore

$$X = \begin{pmatrix} 3 & -4 \\ -1 & 3/2 \end{pmatrix} \times \begin{pmatrix} 6 \\ 2 \end{pmatrix}$$
$$= \begin{pmatrix} (3 \times 6) + (-4 \times 2) \\ (-1 \times 6) + (3/2 \times 2) \end{pmatrix}$$

$$= \begin{pmatrix} 10 \\ -3 \end{pmatrix}$$

7.7 Application to Computing

Matrices are particularly important in the area of computer graphics. A digital image is essentially a matrix. The rows and columns of the matrix correspond to rows and columns of pixels. The numerical value of each entry corresponds to the colour value of the pixel. Matrix algebra is extremely important in performing the operations necessary to manipulate digital images. Decoding digital video requires numerous operations such as matrix multiplication, transformations and so on.

7.8 Using Matrices to Solve Linear Equations

Matrices provide a convenient way of solving simultaneous linear equations.

Consider the following three equations:

$$x + y + z = 6$$
$$2y + 5z = -4$$
$$2x + 5y - z = 27$$

We can form a matrix from the coefficients of the variables:

$$\begin{pmatrix} 1 & 1 & 1 \\ 0 & 2 & 5 \\ 2 & 5 & -1 \end{pmatrix}$$

And we can form another matrix from the variables:

$$\begin{pmatrix} x \\ y \\ z \end{pmatrix}$$

From our knowledge of matrix multiplication we have:

$$\begin{pmatrix} 1 & 1 & 1 \\ 0 & 2 & 5 \\ 2 & 5 & -1 \end{pmatrix} \begin{pmatrix} x \\ y \\ z \end{pmatrix} = \begin{pmatrix} x + y + z \\ 2y + 5z \\ 2x + 5y - z \end{pmatrix}$$

We will refer to these three matrices as A, X and B respectively.

7.8 Using Matrices to Solve Linear Equations

If we substitute the values of the right hand side of our equations into B we have:

$$\begin{pmatrix} 1 & 1 & 1 \\ 0 & 2 & 5 \\ 2 & 5 & -1 \end{pmatrix} \begin{pmatrix} x \\ y \\ z \end{pmatrix} = \begin{pmatrix} 6 \\ -4 \\ 27 \end{pmatrix}$$

So any set of linear equations can be written in the form: $AX = B$.
In our example:

$$x + y + z = 6$$
$$2y + 5z = -4$$
$$2x + 5y - z = 27$$

Writing these in the form $AX = B$:

$$A = \begin{pmatrix} 1 & 1 & 1 \\ 0 & 2 & 5 \\ 2 & 5 & -1 \end{pmatrix} \quad X = \begin{pmatrix} x \\ y \\ z \end{pmatrix} \quad B = \begin{pmatrix} 6 \\ -4 \\ 27 \end{pmatrix}$$

To solve our equations we need to find X, which will give us the values for x, y, and z. We use the fact that:

$$\text{If: } AX = B \quad \text{Then: } X = A^{-1}B.$$

Using a calculator we find that

$$A^{-1} = \begin{pmatrix} \frac{9}{7} & \frac{-2}{7} & \frac{-1}{7} \\ \frac{-10}{21} & \frac{1}{7} & \frac{5}{21} \\ \frac{4}{21} & \frac{1}{7} & \frac{-2}{21} \end{pmatrix}$$

Again using a calculator:

$$X = A^{-1}B = \begin{pmatrix} 5 \\ 3 \\ -2 \end{pmatrix}$$

So the solution to our equation is: $x = 5$, $y = 3$, $z = -2$.

7.9 Row Operations on a Matrix

There are a number of operations that can be performed on the rows of a matrix which prove useful in other calculations.

1. Switch any two rows:

$$\begin{pmatrix} 2 & 1 & 5 \\ 4 & 3 & 6 \\ 6 & 7 & 8 \end{pmatrix} \rightarrow \begin{pmatrix} 4 & 3 & 6 \\ 2 & 1 & 5 \\ 6 & 7 & 8 \end{pmatrix}$$

$$R_1 \leftrightarrow R_2.$$

2. Multiply a row by a nonzero constant:

$$\begin{pmatrix} 2 & 1 & 5 \\ 4 & 3 & 6 \\ 6 & 7 & 8 \end{pmatrix} \rightarrow \begin{pmatrix} 4 & 2 & 10 \\ 4 & 3 & 6 \\ 6 & 7 & 8 \end{pmatrix}$$

$$2R_1 \rightarrow R_1$$

3. Add one row to another and replace one of the rows:

$$\begin{pmatrix} 2 & 1 & 5 \\ 4 & 3 & 6 \\ 6 & 7 & 8 \end{pmatrix} \rightarrow \begin{pmatrix} 2 & 1 & 5 \\ 6 & 4 & 11 \\ 6 & 7 & 8 \end{pmatrix}$$

$$R_1 + R_2 \rightarrow R_2.$$

Operations 2 and 3 can be combined:

$$\begin{pmatrix} 2 & 1 & 5 \\ 4 & 3 & 6 \\ 6 & 7 & 8 \end{pmatrix} \rightarrow \begin{pmatrix} 10 & 7 & 17 \\ 4 & 3 & 6 \\ 6 & 7 & 8 \end{pmatrix}$$

$$R_1 + 2R_2 \rightarrow R_1.$$

Note: Multiplying one row by -1 and adding is equivalent to subtracting one row from another.

7.10 Solving Equations Using the Gauss-Jordan Elimination Method

This method of solving equations is not easy and requires some patience and intuition. It is unlikely that you would be required to perform this process in the early stages of a computer science course, but it is nonetheless something you need to be aware of, as it is an important tool in more advanced aspects of computing such as *image and graphics processing,* specifically in transforming, scaling, and rotating images.

1. Write the **augmented matrix** of the system of equations.

For example, for the following set of equations:

$$x + y + z = 5$$
$$2x + 3y + 5z = 8$$
$$4x + 5z = 2$$

the augmented matrix is:

$$\begin{pmatrix} 1 & 1 & 1 & | & 5 \\ 2 & 3 & 5 & | & 8 \\ 4 & 0 & 5 & | & 2 \end{pmatrix}$$

2. Perform row operations on the augmented matrix until the matrix on the left hand side of the divider is an identity matrix.
3. Read the solution from the right hand side.

Example

$$x + y + z = 5$$
$$2x + 3y + 5z = 8$$
$$4x + 5z = 2$$

1. Write the augmented matrix:

$$\begin{pmatrix} 1 & 1 & 1 & | & 5 \\ 2 & 3 & 5 & | & 8 \\ 4 & 0 & 5 & | & 2 \end{pmatrix}$$

2. Perform row operations:

$$R_2 - 2R_1 \to R_2 \begin{pmatrix} 1 & 1 & 1 & | & 5 \\ 0 & 1 & 3 & | & -2 \\ 4 & 0 & 5 & | & 2 \end{pmatrix}$$

$$R_3 - 4R_1 \to R_3 \begin{pmatrix} 1 & 1 & 1 & | & 5 \\ 0 & 1 & 3 & | & -2 \\ 0 & -4 & 1 & | & -18 \end{pmatrix}$$

$$R_3 + 4R_2 \to R_3 \begin{pmatrix} 1 & 1 & 1 & | & 5 \\ 0 & 1 & 3 & | & -2 \\ 0 & 0 & 13 & | & -26 \end{pmatrix}$$

$$\frac{1}{13} R_3 \begin{pmatrix} 1 & 1 & 1 & | & 5 \\ 0 & 1 & 3 & | & -2 \\ 0 & 0 & 1 & | & -2 \end{pmatrix}$$

$$R_2 - 3R_3 \to R_2 \begin{pmatrix} 1 & 1 & 1 & | & 5 \\ 0 & 1 & 0 & | & 4 \\ 0 & 0 & 1 & | & -2 \end{pmatrix}$$

$$R_1 - R_3 \to R_1 \begin{pmatrix} 1 & 1 & 0 & | & 7 \\ 0 & 1 & 0 & | & 4 \\ 0 & 0 & 1 & | & -2 \end{pmatrix}$$

$$R_1 - R_2 \to R_1 \begin{pmatrix} 1 & 0 & 0 & | & 3 \\ 0 & 1 & 0 & | & 4 \\ 0 & 0 & 1 & | & -2 \end{pmatrix}$$

3. The solution to the equation is:

$$\boxed{x = 3 \quad y = 4 \quad z = -2}$$

7.11 Exercises

1. Consider the following matrices:

$$A = \begin{pmatrix} -2 & 3 & 5 \\ 1 & -2 & 9 \end{pmatrix} \quad B = \begin{pmatrix} 1 & 2 & 7 \\ 3 & 8 & 4 \end{pmatrix}$$

7.11 Exercises

Find the value of:

(a) $A + B$ (b) $A - B$ (c) $2A + 3B$ (d) A^T

2. Consider the following matrices:

$$A = (-2 \ 3) \quad B = \begin{pmatrix} 4 & 1 \\ 2 & 3 \end{pmatrix}$$

Calculate $A \times B$.

3. Consider the following matrices:

$$A = \begin{pmatrix} 2 & 7 \\ 1 & 3 \end{pmatrix} \quad B = \begin{pmatrix} 1 & 5 \\ 0 & 2 \end{pmatrix}$$

Calculate $A \times B$.

4. Find the determinant of the following matrix, A:

$$A = \begin{pmatrix} 1 & 3 \\ 4 & -2 \end{pmatrix}$$

5. Where possible, find the inverse of the following matrices:

(a) $A = \begin{pmatrix} 4 & 2 \\ 1 & 3 \end{pmatrix}$ (b) $B = \begin{pmatrix} 1 & 4 \\ 2 & 5 \\ 3 & 6 \end{pmatrix}$ (c) $C = \begin{pmatrix} 3 & 6 \\ 2 & 4 \end{pmatrix}$

6. Consider the matrix A from the previous question:

$$A = \begin{pmatrix} 4 & 2 \\ 1 & 3 \end{pmatrix}$$

Now consider the following matrix D.

$$D = \begin{pmatrix} 2 \\ 2 \end{pmatrix}$$

If $A \times X = D$, find the value of X.

7. Use matrices to solve the following equations (you can use an online matrix calculator to find the inverse and perform the matrix multiplication):

$$2x + y + 4z = 7$$
$$5y + z = 3$$
$$-x + 4y + 2z = -2$$

8. Use the Gauss-Jordan elimination method to solve the following simultaneous equations:

$$2x + 5y = 21$$
$$x + 2y = 8$$

Combinatorics

At the end of this chapter you should be able to:

- explain the concept of a **factorial** and perform calculations on factorials;
- distinguish between a **permutation** and a **combination**;
- make calculations involving permutations and combinations, with and without repetition;
- analyse a particular scenario to determine the correct formula to use;
- explain how **Pascal's Triangle** is formed;
- describe how Pascal's Triangle is related to the subject of combinations;
- explain the importance of combinatorics to computer science.

8.1 Introduction

The main topics introduced in this chapter, **permutations** and **combinations**, form part of the overall branch of mathematics known as **combinatorics**. Combinatorics is primarily concerned with counting. We will learn here how to count the number of ways that items can be arranged, and how many ways we can select items from a larger group of items, depending upon whether or not the order in which we select them is significant, and whether or not we are allowed to repeat items. We will later go on to explore how this topic provides a convenient way of expanding binomial algebraic expressions.

Table 8.1 Possible outcomes from a race with 6 runners

Position	Ways of choosing	Comments
1	6	There are 6 different possible ways of choosing the winner
2	6×5	For each winner there are 5 different choices for 2nd place. So there are 6×5 (or 30) ways of choosing the first 2 people
3	$6 \times 5 \times 4$	There are 4 choices for 3rd place. So the number of ways of choosing the first 3 is $6 \times 5 \times 4 = 120$
4	$6 \times 5 \times 4 \times 3$	We continue in the same way. So the number of ways of choosing the first 4 is $6 \times 5 \times 4 \times 3 = 360$
5	$6 \times 5 \times 4 \times 3 \times 2$	The number of ways of choosing the first 5 is $6 \times 5 \times 4 \times 3 \times 2 = 720$
6	$6 \times 5 \times 4 \times 3 \times 2 \times 1$	Once the first 5 are chosen, there is only one choice for last place. So again there are 720 ($6 \times 5 \times 4 \times 3 \times 2 \times 1$) possibilities

8.2 Placing Items in Order

Imagine six runners in a race. How many possible outcomes could there be? In other words how many different ways are there of ordering the six competitors? Table 8.1 shows the possible outcomes.

We see that the number of ways of arranging 6 items in order is: $6 \times 5 \times 4 \times 3 \times 2 \times 1 = 720$.

8.3 Factorials

The sequence that we have just seen ($6 \times 5 \times 4 \times 3 \times 2 \times 1$) is given a special name. It is called 6 factorial and is written with an exclamation mark, like this: 6!

So: $6! = 6 \times 5 \times 4 \times 3 \times 2 \times 1 = 720$
$7! = 7 \times 6 \times 5 \times 4 \times 3 \times 2 \times 1 = 5040$
$4! = 4 \times 3 \times 2 \times 1 = 24$

And so on.

In general: $n! = n \times (n-1) \times (n-2) \times (n-3) \times \ldots \ldots \times 1.$

Worked Example 1

In how many different ways can the letters a, b, c, d be arranged?

Solution

There are 4 letters, so the number of ways of arranging them is

$$4! = 4 \times 3 \times 2 \times 1 = 24.$$

Worked Example 2

Find the value of $\frac{7! \times 4!}{5!}$.

Solution

$$\frac{7! \times 4!}{5!} = \frac{7 \times 6 \times \cancel{5} \times \cancel{4} \times \cancel{3} \times \cancel{2} \times \cancel{1} \times 4 \times 3 \times 2 \times 1}{\cancel{5} \times \cancel{4} \times \cancel{3} \times \cancel{2} \times \cancel{1}}$$
$$= 7 \times 6 \times 4 \times 3 \times 2 \times 1$$
$$= 1008$$

8.4 Choosing just Some of the Items

Consider again the six runners. Imagine we are interested only in the first two places. The number of ways we can select some or all items from a list, when (like in this example) *the order is important* is called a **permutation**

A permutation of 2 items from 6 is written like this: $P(6, 2)$.

Another common notation is: $^6P_2$.

Look again at Table 8.1. You can see that if we are only interested in the first two places (the first and second rows of the table), then instead of calculating the whole sequence, (6 factorial) we use only the first two and discard the last four.

Effectively we divided 6 factorial by 4 factorial:

$$P(6, 2) = \frac{6!}{(6-2)!} = \frac{6!}{4!}$$
$$= \frac{6 \times 5 \times 4 \times 3 \times 2 \times 1}{4 \times 3 \times 2 \times 1}$$
$$= 6 \times 5 = 30$$

Similarly, if we wanted to know the possibilities for the first 4 places:

$$P(6, 4) = \frac{6!}{(6-4)!} = \frac{6!}{2!}$$
$$= \frac{6 \times 5 \times 4 \times 3 \times 2 \times 1}{2 \times 1}$$
$$= 6 \times 5 \times 4 \times 3 = 360$$

8.5 The Formula for Permutations

In general, if we want to know how many ways we can choose k items from n items when the order in which we select them **is** important, we use the following formula:

$$P(n,k) = \frac{n!}{(n-k)!}$$

8.6 Combinations

Imagine a situation, such as a lottery, where we select a number of items, but the order in which they are selected doesn't matter. The number of ways we can select several items from a list, when the order is **not** important, is called a **combination.**

The notation for combinations is similar to permutations, but uses a C instead of a P. So we can write $C(6,2)$ or $^6C_2$.

Another common notation for combinations is: $\binom{n}{k}$.

For example: $\binom{6}{2}$.

Suppose we have a small lottery consisting of 12 balls, and we have to choose three. There will be fewer choices now. When we were dealing with permutations we could choose three items - for example the numbers 5, 6, 7 – and arrange them in 3! ways. But if order is unimportant there is only one way. So we have to divide the permutation by 3 factorial.

$$\begin{aligned} C(12,3) &= \frac{P(12,3)}{3!} = \frac{12!}{9! \times 3!} \\ &= \frac{12 \times 11 \times 10 \times 9 \times 8 \times 7 \times 6 \times 5 \times 4 \times 3 \times 2 \times 1}{9 \times 8 \times 7 \times 6 \times 5 \times 4 \times 3 \times 2 \times 1 \times 3 \times 2 \times 1} \\ &= \frac{12 \times 11 \times 10}{3 \times 2 \times 1} = 220 \end{aligned}$$

8.7 The Formula for Combinations

In general, if we want to know how many ways we can choose k items from n items when the order in which we select them is **not** important, we use the following formula:

$$C(n,k) = \frac{n!}{(n-k)!k!}$$

8.8 What Is the Value of 0!?

There are times when a calculation results in 0!, so this needs to have a value. Imagine we have n objects. In how many different ways can we choose zero of these objects? The answer has to be 1. There is only one way of choosing zero items.

So logically: $C(n, 0) = 1$.

Therefore, using our formula for combinations:

$$\frac{n!}{n! \times 0!} = 1$$

You can see that the only way this can work is if 0! is equal to 1.

8.9 Allowing Repetition (When Order Is Important)

Imagine you have to choose a 4-digit code for an intruder alarm. There are 10 digits available, from 0 to 9. Order is important. However, this time you are allowed to repeat digits—for example a code of 5508 or 4142.

There are 10 ways of choosing the first digit.

There are 10 ways of choosing the second digit—so the number of ways of choosing the first 2 digits is: 10×10.

There are 10 ways of choosing the third digit—so the number of ways of choosing the first 3 digits is: $10 \times 10 \times 10$.

There are 10 ways of choosing the final digit—so the number of ways of choosing all 4 digits is: $10 \times 10 \times 10 \times 10$.

The number of different codes is: $10^4 = 10,000$.

In general, if we want to know how many ways we can choose k items from n items when the order in which we select them **is** important **and repetitions are allowed**, we use the following formula:

$$n^k$$

8.10 Allowing Repetition (When Order Is Not Important)

In certain restaurants you can choose dishes by means of a colour code. Each colour represents a particular price. Imagine there are four colours available – red, blue, green, yellow. You are going to choose three dishes.

You want to know how many different combinations of dishes you could have - each colour can be chosen more than once. All the possible combinations are listed below:

Table 8.2 Choosing k items from n items

	Repetitions NOT allowed	Repetitions allowed
Order IS important (Permutation)	$P(n, k) = \frac{n!}{(n-k)!}$	$P(n, k) = n^k$
Order IS NOT important (Combination)	$C(n, k) = \frac{n!}{(n-k)!k!}$	$C(n, k) = \frac{(n+k-1)!}{k!(n-k)!}$

(R, R, R)	(R, B, B)	(R, G, Y)	(B, B, Y)	(G, G, G)
(R, R, B)	(R, B, Y)	(R, Y, Y)	(B, G, G)	(G, G, Y)
(R, R, G)	(R, G, Y)	(B, B, B)	(B, G, Y)	(G, Y, Y)
(R, R, Y)	(R, G, G)	(B, B, G)	(B, Y, Y)	(Y, Y, Y)

There are 20 different possibilities. The formula to use in this situation is given below. It is an easy formula to use, but is harder to derive, so we will show you the formula now, and explain how this is arrived at later.

In general, if we want to know how many ways we can choose k items from n items when the order in which we select them **is not** important **and repetitions are allowed**, we use the following formula:

$$\frac{(n+k-1)!}{k!(n-1)!}$$

8.11 Summary of Formulae

The different ways of selecting k items from n items are shown in Table 8.2.

8.12 Deriving the Formula for Choosing Items when Order Is Not Important and Repetition Is Allowed

We will refer to the previous example of the restaurant with colour-coded dishes. This time we will have five colours—red, blue, green, yellow and pink. We will again select three dishes.

Since order is not important, we have to find a way of eliminating duplicates such as (R, R, G) and (R, G, R). Because we are allowing repetitions, it is more complicated than simply dividing the permutation by 3! like we did before.

One way to do it is to go through the colours in order, place a marker to show that an item has been selected, and then place a different marker between each type of item.

8.12 Deriving the Formula for Choosing Items when Order Is Not ...

We will list the items in this order: R, B, G, Y, P and we will use 1 to show an item is selected and 0 to separate the items. An example will make this clear.

$$(R, R, G) \text{ would be } (1\,1\,0\,0\,1\,0\,0)$$

Here we have put a 1 for red, another 1 for red, then a zero to show we have moved to blue. There are no blues, so another 0 shows we have moved to green. We have placed a 1 for green, then another zero to show we have moved to yellow. Finally another zero to show we have moved to pink.

$$(R, G, R) \text{ would be the same.}$$

So we have found a way to have only one string for each combination. Another couple of examples:

$$(R, Y, P) \text{ would be } (1 0\,0\,0 1 0 1)$$

$$(B, Y, Y) \text{ would be } (0 1 0\,0 1 1 0)$$

You can see that there are always four 0s and three 1s; seven places in all.

Our question can now be put in a different way: If there are 7 places, how can we choose 3 of them to put our 1s in? We are finding $C(7, 3)$.

In general: There will always be one fewer 0's than the total number of items, as there is a 0 between each item. So if there are n items and we are choosing k of them, then there will be $n + k - 1$ places, and we must chose k of them. We need to calculate $C(n + k - 1, k)$.

So we have:

$$C(n + k - 1, k) = \frac{(n+k-1)!}{k!(n+k-1-k)!}$$
$$= \frac{(n+k-1)!}{k!(n-1)!}$$

Worked Example 3
A teacher has a class of 30 students. She has to chose 3 students to be awarded a prize for good progress.

(a) How many different sets of three can be chosen?

Solution
As there is no ordering involved here, and no repetition, we use the formula for combinations without repetition:

$$C(n, k) = \frac{n!}{(n-k)!k!}$$

In this case $n = 30$ and $k = 3$.

$$C(30, 3) = \frac{30!}{27! \times 3!}$$
$$= \frac{30 \times 29 \times 28}{3 \times 2 \times 1}$$
$$= 4060$$

(b) The teacher now has to choose 3 students to be awarded 1st, 2nd and 3rd prize for overall performance. How many different possibilities are there?

Solution
This time there is ordering, but once again no repetition, so we use the formula for permutations without repetition:

$$P(n, k) = \frac{n!}{(n-k)!}$$

As in the previous question, $n = 30$ and $k = 3$

$$P(30, 3) = \frac{30!}{27!}$$
$$= 30 \times 29 \times 28$$
$$= 24360$$

(c) A prize is to be awarded for mathematics, science, and languages. No student should receive more than one prize. How many different possibilities are there?

Solution
In fact this is exactly the same as part b). Just think of it as placing the students in a line with, say, the winners of the mathematics, science and language prizes in first, second and third position.
So the answer is the same as before, 24360

(d) Once again there are to be three awards—mathematics, science and languages. But this time the rules are relaxed, and each student is eligible for all three prizes. So any student could receive as many as three prizes. How many possibilities are there?

Solution
Once again there is ordering, but this time with repetition, so the correct calculation is:

$$n^k = 30^3 = 27000$$

Worked Example 4
A five-a-side football club consists of six different teams. Three members of the club are to be chosen for a special award. They can be chosen from any team, so that a particular team could end up winning three awards, two awards, one award or no awards. In how many different ways can the awards be distributed among the teams?

Solution
There is no ordering but there is repetition, because one team could win all three awards, none of the awards, or any number in between.

Because there are six teams, $n = 6$ and $k = 3$.

The correct calculation for no ordering with repetition allowed is:

$$\frac{(n+k-1)!}{k!(n-1)!} = \frac{8!}{3! \times 5!}$$
$$= \frac{8 \times 7 \times 6}{3 \times 2 \times 1}$$
$$= 56$$

Worked Example 5
Five friends are planning a trip to the cinema. They will all sit together in a row. Leroy and Calvin want to sit together. In how many different ways can the five friends occupy the row?

Solution
The easiest thing to do is to consider Leroy and Calvin as one item. We now have to work out how many ways to arrange four items, which is 4!.

$$4! = 4 \times 3 \times 2 \times 1 = 24.$$

There are, however, 2! ways in which Leroy and Calvin can be arranged, so for each of the above arrangements there are 2 options.

So the total number of arrangements is: $2 \times 24 = 48$.

Worked Example 6
How many 4-digit numbers can be made from the digits 1–9 if you are not allowed to repeat numbers, and the number must be divisible by 5?

Solution

If the number is divisible by 5, it must end in a 5 (since there is no zero). So this comes down to arranging 3 numbers out of 8, without repetition.

$$P(8, 3) = \frac{8!}{5!}$$
$$= 8 \times 7 \times 6$$
$$= 336$$

8.13 Pascal's Triangle

Pascal's triangle is a very well-known number pattern, used in several areas of mathematics. The first 8 rows are shown in Fig. 8.1.

Each number is made up by adding the two numbers above it (shown in the diagram by the red and green squares).

Interestingly, each number represents $C(n, k)$, where n is the row number (starting from 0) and k the column number (also starting from 0). For example, the blue square containing the number 15, represents $C(6, 2)$ and the yellow square, containing the number 21, is $C(7, 5)$.

Pascal's triangle demonstrates clearly that combinations are symmetrical. $C(n, k)$ is the same as $C(n, n-k)$. For example, $C(8, 3)$ is the same as $C(8, 5)$. This can also be seen from the formula for $C(n, k)$—and by logic: choosing 2 items from 7 is the same as choosing 5 items to discard.

Worked Example 7

(a) Use Pascal's Triangle to find the value of $C(5,2)$.
(b) Verify your answer using the formula.

Fig. 8.1 Pascal's triangle

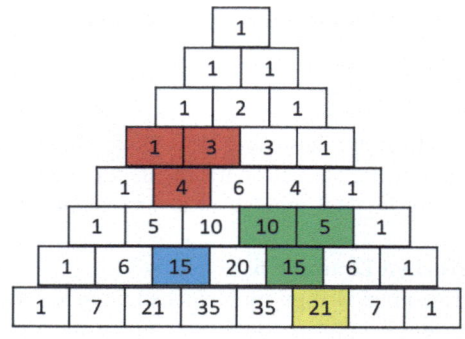

8.14 Binomial Expansion

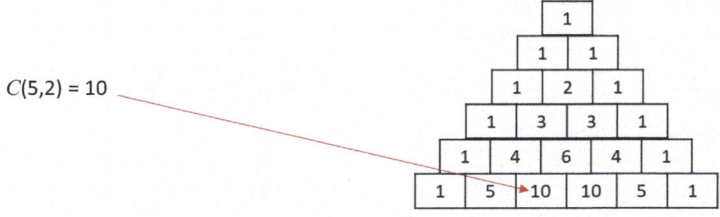

Fig. 8.2 Solution to worked example 7

Solution

(a) Because we start at zero, we will need to draw 6 rows of the triangle as shown in Fig. 8.2

(b) $C(5,2) = \frac{5!}{3! \times 2!} = \frac{5 \times 4 \times 3 \times 2 \times 1}{3 \times 2 \times 1 \times 2 \times 1} = 10$

8.14 Binomial Expansion

A binomial expression is an expression that contains two terms.

Imagine we had to calculate an expression such as: $(a + b)^n$ where n is a natural number. Our knowledge of combinatorics can help us with this. The first few expansions would give us:

$$(a + b)^0 = 1$$
$$(a + b)^1 = a + b$$
$$(a + b)^2 = a^2 + 2ab + b^2$$
$$(a + b)^3 = a^3 + 3a^2b + 3ab^2 + b^3$$
$$(a + b)^4 = a^4 + 4a^3b + 6a^2b^2 + 4ab^3 + b^4$$

The coefficients form the following pattern:

$$\begin{array}{ccccccccc} & & & & 1 & & & & \\ & & & 1 & & 1 & & & \\ & & 1 & & 2 & & 1 & & \\ & 1 & & 3 & & 3 & & 1 & \\ 1 & & 4 & & 6 & & 4 & & 1 \end{array}$$

which you will recognize as Pascal's triangle.

Knowing, as we do, that each entry in Pascal's triangle represents a combination, we can rewrite our expansions.

For example we saw that:
$(a+b)^4 = a^4 + 4a^3b + 6a^2b^2 + 4ab^3 + b^4$.
We could write this as:
$(a+b)^4 = {}^4C_0 a^4 b^0 + {}^4C_1 a^3 b^1 + {}^4C_2 a^2 b^2 + {}^4C_3 a^1 b^3 + {}^4C_4 a^0 b^4$.
In general we could write:
$(a+b)^n = {}^nC_0 a^n b^0 + {}^nC_1 a^{n-1} b^1 + {}^nC_2 a^{n-2} b^2 + \cdots + {}^nC_{n-1} a^1 b^{n-1} + {}^nC_n a^0 b^n$.
In mathematics, we often use the upper case Greek letter sigma, $\sum$, to mean *the sum of*. Using this notation we could write:

$$(a+b)^n = \sum_{k=0}^{n} {}^nC_k a^{n-k} b^k$$

This means that you start of with $k=0$, than add every term until you reach n. This is known as the **binomial theorem**.

Worked Example 8
Use the binomial theorem to expand the expression $(x + 2y)^5$

Solution

$$(a+b)^n = {}^nC_0 a^n b^0 + {}^nC_1 a^{n-1} b^1 + {}^nC_2 a^{n-2} b^2 + \ldots + {}^nC_{n-1} a^1 b^{n-1} + {}^nC_n a^0 b^n$$

The first term in our expression is x, the second is $2y$, and in this case $n=5$.

$(x+2y)^5 = {}^5C_0 x^5 (2y)^0 + {}^5C_1 x^4 (2y)^1 + {}^5C_2 x^3 (2y)^2 + {}^5C_3 x^2 (2y)^3 + {}^5C_4 x^1 (2y)^4 + {}^5C_5 x^0 (2y)^5$
$= x^5 + 5x^4(2y) + 10x^3(2y)^2 + 10x^2(2y)^3 + 5x(2y)^4 + (2y)^5$
$= x^5 + 10x^4 y + 40x^3 y^2 + 80x^2 y + 80xy^4 + 32y^5$

Worked Example 9
Use the binomial theorem to find the 3rd term in the expansion of the expression $(x-y)^7$

Solution
We saw that

$$(a+b)^n = \sum_{k=0}^{n} {}^nC_k a^{n-k} b^k$$

Because we start at 0, the 3rd term will be found when $k=2$.
Here $a=x$ and $b=-y$, and $n=7$.
The third term is therefore: ${}^7C_2 x^5 (-y)^2 = 21 x^5 y^2$

8.15 Application to Computing

There are many examples in computer science where we need to compute the number of different ways of doing things.

In computer graphics, for example, colour is made by combining the three primary colours, red, blue and green, in different intensities. Different systems are used to achieve this, and it is very important to know how many different colours can be obtained using a particular system.

Another very important application is in the area of **algorithm efficiency**. When writing the code for a particular task, there are always many ways of achieving the same goal. Some algorithms are a great deal more efficient than others, because they require fewer steps. Combinatorics plays a very important role in comparing different algorithms by determining the number of possible steps involved in each one.

8.16 Exercises

1. In how many different ways can the letters x, y, z, w, v be arranged?
2. Find the value of:

$$\frac{8! \times 5!}{4! \times 3!}$$

3. Find the value of: (a) $P(10, 3)$ (b) $C(9, 6)$.
4. A committee of 20 people has to elect a chair, a vice-chair, a secretary and a treasurer. How many different ways are there of choosing these posts?
5. The winner of a children's competition is allowed to draw three prizes from a bag of 10 unique items. The runner up is then allowed to draw two items.

   How many different sets of prizes can be chosen by:
   (a) the winner;
   (b) the runner-up?
6. A gift shop sells 10 different colours of wrapping paper. Customers can get a discount if they buy three rolls of paper. They can choose three of the same colour, or two of one colour and one of another colour, or 3 different colours.

   How many different combinations can a customer choose from?
7. Imagine an alien alphabet consisting of the following symbols:

$$€ \quad \% \quad @ \quad \pi \quad * \quad £$$

   How many different three letter "words" can be made from these symbols? (Symbols can be repeated).
8. How many 3-digit numbers can be made from the digits 1–6 if:
   (a) you are allowed to repeat digits;
   (b) you are not allowed to repeat digits;
   (c) you are not allowed to repeat digits and the number must end in 3;
   (d) you are not allowed to repeat digits and the number must end in 1 or 4?

9. Four friends go on a fairground ride. They must sit in a row. Tracey does not want to sit at the end of the row. In how many different ways can the four be arranged?
10. Use Pascal's triangle to find the value of $C(4, 2)$; verify your answer by using the correct formula.
11. Use the binomial theorem to expand the expression $(2x - y)^4$
12. Use the binomial theorem to find the 3rd term in the expansion of the expression $(x + 2y)^6$

Probability

At the end of this chapter you should be able to:

- define the terms **experiment**, **outcome**, **sample space** and **event**;
- determine distinct sets for a sample space and event from a given scenario;
- provide a definition of **probability**;
- calculate the probability of a particular event being successful;
- represent a probability distribution using a **discrete random variable**;
- calculate the **expected value** of an experiment;
- distinguish between **mutually exclusive** and **non-mutually exclusive** events;
- use the **addition rule** to determine the probability of two alternative events happening;
- distinguish between independent and conditional events, and calculate the **probability** in both cases;
- construct and interpret **tree diagrams** for conditional events;
- explain and apply **Bayes' theorem**;
- make use of **the binomial probability formula**.

9.1 Introduction

In this chapter you will be studying **probability**, which is the branch of mathematics that builds a framework around random, or chance, events. Probability plays an extremely important role in many other disciplines such as physics, biology, finance, and, as we shall see, in computer science.

Fig. 9.1 A 5-sided spinner

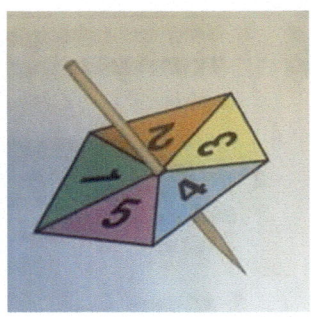

9.2 Terminology and Definitions

Consider the 5-sided spinner that is shown in Fig. 9.1.

If the spinner is not weighted in any way, and there is an equal chance of it landing on any one of the five numbers, it is called a **fair** spinner.

If we spin the spinner once, this is called an **experiment**. The number that it lands on is called an **outcome**.

The set of all possible outcomes is called the **sample space**. A set of possible outcomes is called an **event**.

The **probability** of an event is a measure of how likely it is that that event will happen.

9.3 Outcomes, Sample Space and Events

For our spinner there are five possible outcomes:

The spinner can land on 1.
The spinner can land on 2.
The spinner can land on 3.
The spinner can land on 4.
The spinner can land on 5.

If we call the sample space for the spinner S, then:

$$S = \{1, 2, 3, 4, 5\}$$

Let's define a few events:

A: the event that the spinner lands on a 5
B: the event that the spinner lands on a 3 or 4
C: the event that thes pinner lands on an odd number

D: the event that the spinner lands on a number less than 5.

We see that:

$$A = \{5\} \quad B = \{3, 4\} \quad C = \{1, 3, 5\} \quad D = \{1, 2, 3, 4\}$$

9.4 Calculating Probability

The probability that an event will happen is given by the following formula:

$$probability = \frac{number\ of\ outcomes\ in\ the\ event}{number\ of\ outcomes\ in\ the\ sample\ space}$$

We use the notation $P(E)$ to mean the probability that E will happen. We can also use the notation for cardinality that we learnt when studied set theory:

If E is an event and S is the sample space, then:

$$P(E) = \frac{n(E)}{n(S)}$$

For the examples on the previous slide:

$$S = \{1, 2, 3, 4, 5\}$$

Therefore

$$n(S) = 5$$

Table 9.1 shows how the probability of each of the events described in the previous section is calculated.

We can express the probability as a fraction, a decimal or a percentage. So we could have written, for example, $P(B) = 0.4$ or $P(B) = 40\%$.

It is worth noting that:

- If an event is certain to happen as a result of a particular experiment the probability is 1. For example, the probability of our spinner landing on either a 1, 2, 3, 4 or 5 is 1.

Table 9.1 Calculating probabilities

Event	$A = \{5\}$	$B = \{3, 4\}$	$C = \{1, 3, 5\}$	$D = \{1, 2, 3, 4\}$
Cardinality	$n(A) = 1$	$n(B) = 2$	$n(C) = 3$	$n(D) = 4$
Probability	$P(A) = 1/5$	$P(B) = 2/5$	$P(B) = 3/5$	$P(D) = 4/5$

- If an event is impossible the probability is 0. For example, the probability of our spinner landing on a 6 is 0.
- All other probabilities will lie between 0 and 1.

Worked Example 1
A 6-sided dice is thrown.

(a) What is the sample space, S?
(b) Give the value of E, the event that the dice lands on a number less than 3.
(c) What is the cardinality of the set S and the set E?
(d) Calculate the probability that E happens.

Solution

(a) $S = \{1, 2, 3, 4, 5, 6\}$
(b) $E = \{1, 2\}$
(c) $n(S) = 6 \quad n(E) = 2$
(d) $P(E) = \frac{n(E)}{n(S)} = \frac{1}{3}$.

Worked Example 2
Two 3-sided spinners (see Fig. 9.2) are spun together.

(a) What is the sample space, S?
(b) Give the value of E, the event one spinner lands on blue and the other on yellow.
(c) What is the cardinality of the set S and the set E?
(d) Calculate the probability that E happens.

Fig. 9.2 Two 3-sided spinners

9.6 Mutually Exclusive Events

Solution

(a)
$S = \{$(RED, RED), (RED, BLUE), (RED, YELLOW),
(BLUE, RED), (BLUE, BLUE), (BLUE, YELLOW),
(YELLOW, RED), (YELLOW, BLUE), (YELLOW, YELLOW)$\}$.

(b) $E = \{$(BLUE, YELLOW), (YELLOW, BLUE)$\}$.

(c) $n(S) = 9 \quad n(E) = 2$.

(d) $P(E) = \frac{n(E)}{n(S)} = \frac{2}{9}$.

9.5 Probability that One or Another Event Happens

Consider the 5-sided spinner again. Let A be the event of getting a 2 and B be the event of getting a 5:

$$A = \{2\} \quad B = \{5\}$$

The event of getting a 2 *or* a 5 is: $\{2, 5\}$.

We know that this is the union of A and B. So the event of getting a 2 *or* a 5 is $A \cup B$. So we see that the probability of event A *or* B happening is:

$$P(A \cup B)$$

9.6 Mutually Exclusive Events

The examples we have just seen are **mutually exclusive** events. They cannot both happen at the same time; a spinner cannot land on a 2 and a 5 at the same time. Of course, events containing a single outcome are always mutually exclusive.

When the events contain more than one outcome, they may or may not be mutually exclusive.

For example, if you throw a dice, the event of getting an even number and the event of getting a 5 *are* mutually exclusive. For mutually exclusive events, there are no common elements.

In other words, for two mutually exclusive events, A and B:

$$A \cap B = \emptyset$$

Consider throwing a 6-sided dice:

Let A be the event of throwing an odd number.
Let B be the event of throwing a number less than 4.

$$A = \{1, 3, 5\} \quad B = \{1, 2, 3\}$$

These events are *not* mutually exclusive. It is possible to throw a number which is odd *and* is less than 4 (the number 1 or the number 3). The intersection of the two sets is not empty:

$$A \cap B = \{1, 3\}$$

9.6.1 The Addition Rule

We have seen that the probability of A or B happening is given by $P(A \cup B)$.
In the previous example of the 6-sided dice we had:

A: The event of throwing an odd number.
B: The event of throwing a number less than 4.

$$A = \{1, 3, 5\} \quad B = \{1, 2, 3\} \quad A \cup B = \{1, 2, 3, 5\}$$

so

$$P(A \cup B) = \frac{4}{6} = \frac{2}{3}$$

There is another way we can calculate the probability of one or another event happening. When we studied sets we saw that:

$$n(A \cup B) = n(A) + n(B) - n(A \cap B)$$

If the sample space is S, then

$$\frac{n(A \cup B)}{n(S)} = \frac{n(A)}{n(S)} + \frac{n(B)}{n(S)} - \frac{n(A \cap B)}{n(S)}$$

So we have:

$$\boldsymbol{P(A \cup B) = P(A) + P(B) - P(A \cap B)}$$

This is called the **addition rule**. In this example:

$$A \cap B = \{1, 3\}$$

So:

$$P(A) = \frac{3}{6} = \frac{1}{2}$$

$$P(B) = \frac{3}{6} = \frac{1}{2}$$

$$P(A \cap B) = \frac{2}{6} = \frac{1}{3}$$

$$P(A \cup B) = \frac{1}{2} + \frac{1}{2} - \frac{1}{3} = \frac{2}{3}$$

This is, of course, the same answer that we got before.

9.6.2 The Addition Rule for Mutually Exclusive Events

For mutually exclusive events there are no common elements:

$$A \cap B = \emptyset$$

So:

$$P(A \cap B) = 0$$

So: *For mutually exclusive events*: $P(A \cup B) = P(A) + P(B)$.
For mutually exclusive events, it is also true that:

- The sum of the individual probabilities must add up to 1.
- The probability of *not* getting a successful event is found by subtracting the probability from 1. For example, the probability of *not* throwing a number below 5 on the spinner is $\left(1 - \frac{1}{5}\right)$ or $\frac{4}{5}$.

Worked Example 3

1. A 6-sided dice is thrown, and the following events are defined:
    A is the event of throwing a number less then 4.
    B is the event of throwing an even number.
    C is the event of throwing a number starting with "F".
   (a) What is the sample space, S?
   (b) Give the value of A, B and C.
   (c) Give the values of: $A \cap B$ $A \cap C$ $B \cap C$.
   (d) Which pairs of events are mutually exclusive?
   (e) Calculate the probability of throwing an even number or a number less than 4.
   (f) Calculate the probability of throwing an even number or a number starting with "F".
   (g) Calculate the probability of throwing a number less than 4 or a number starting with "F".

Solution

(a) $S = \{1, 2, 3, 4, 5, 6\}$.
(b) $A = \{1, 2, 3\}$ $B = \{2, 4, 6\}$ $C = \{4, 5\}$.
(c) $A \cap B = \{2\}$ $A \cap C = \emptyset$ $B \cap C = \{4\}$.
(d) Only A and C (the intersection is empty).
(e) $P(A \cup B) = P(A) + P(B) - P(A \cap B) = \frac{3}{6} + \frac{3}{6} - \frac{1}{6} = \frac{5}{6}$.
(f) $P(B \cup C) = P(B) + P(C) - P(B \cap C) = \frac{3}{6} + \frac{2}{6} - \frac{1}{6} = \frac{2}{3}$.
(g) $P(A \cup C) = P(A) + P(C) - P(A \cap C) = \frac{3}{6} + \frac{2}{6} - 0 = \frac{5}{6}$.

Worked Example 4
Consider the following reduced pack of cards consisting of just the following:

$2\clubsuit, Q\clubsuit$
$4\diamondsuit, 10\diamondsuit, Q\diamondsuit$
$2\heartsuit, Q\heartsuit$
$2\spadesuit, 4\spadesuit, K\spadesuit$

What is the probability of drawing a card that is red or is a queen?

Solution
Let A be the event of drawing a queen. Let B be the event of drawing a red card.

These events are *not* mutually exclusive. You can draw a card that is a queen *and* a red card.

There are 10 cards altogether; there are 5 red cards, there are 3 queens, and there are 2 cards that are red and are queens.

$$P(A) = \frac{3}{10} \quad P(B) = \frac{5}{10} \quad P(A \cap B) = \frac{2}{10}$$

$$P(A \cup B) = \frac{3}{10} + \frac{5}{10} - \frac{2}{10} = \frac{3}{5}$$

9.7 Probability Distribution

When we throw a dice, there are 6 possible outcomes. If the dice is fair, then the probability of throwing any of the numbers from 1 to 6 is $\frac{1}{6}$.

The notation we use for the probability of an event can also be used for the probability of outcomes. So for example P(1) is the probability of throwing a 1.

In general, if A is the event of getting outcome a, then $P(A) = P(a)$.

We write the **probability distribution** for the dice as follows:

$$P(1) = \frac{1}{6}$$

$$P(2) = \frac{1}{6}$$
$$P(3) = \frac{1}{6}$$
$$P(4) = \frac{1}{6}$$
$$P(5) = \frac{1}{6}$$
$$P(6) = \frac{1}{6}$$

Since individual outcomes are mutually exclusive, the total of all the probabilities must be 1. And again because they are mutually exclusive we can add the probabilities to find the probability of one or the other event happening. The probability of throwing a 2 or a 5 is:

$$\frac{1}{6} + \frac{1}{6} = \frac{1}{3}$$

9.7.1 Non-uniform Probability Distribution

Now imagine we toss a coin. This time the coin isn't fair—it has been weighted so that it is 3 times more likely to land on tails than on heads. How can we find the probability distribution?

We let x represent the probability of getting heads. Therefore the probability of getting tails is $3x$.

$$\text{So } P(\text{HEADS}) = x$$
$$P(\text{TAILS}) = 3x$$

We know that the probabilities must add up to 1.
Therefore:

$$x + 3x = 1$$
$$4x = 1$$
$$x = 0.25$$

So our probability distribution is:

$$P(\text{HEADS}) = 0.25$$
$$P(\text{TAILS}) = 0.75$$

Worked Example 5
An unfair 6-sided dice is weighted so that the chances of throwing a 4, 5 or 6 are equal. The chance of throwing a 2 or a 3 is twice as likely as throwing a 4, 5 or 6. The chance of throwing a 1 is three times as likely as throwing a 4, 5 or 6.

(a) Find the probability distribution of the dice.
(b) What is the probability of throwing a 3 or 4?

Solution

(a) Let the probability of throwing a 4, 5, 6 be x. The probability of throwing a 2 or 3 is $2x$. The probability of throwing a 1 is $3x$.

The probability distribution is:

$$P(1) = 3x$$
$$P(2) = 2x$$
$$P(3) = 2x$$
$$P(4) = x$$
$$P(5) = x$$
$$P(6) = x$$

The events are mutually exclusive, so the probabilities must add up to 1:

$$3x + 2x + 2x + x + x + x = 1$$

$$10x = 1 \quad x = 0.1$$

The probability distribution is:

$$P(1) = 0.3$$
$$P(2) = 0.2$$
$$P(3) = 0.2$$
$$P(4) = 0.1$$
$$P(5) = 0.1$$
$$P(6) = 0.1$$

(b) The events are mutually exclusive, therefore the probability of throwing a 3 or a 4 is: $P(3) + P(4) = 0.3$.

9.8 Independent Events

Consider the spinner and the coin shown in Fig. 9.3.

Any events associated with the spinner and the coin are **independent events**. Spinning the spinner has no effect on tossing the coin, and vice versa.

Imagine we want to calculate the probability of getting yellow on the spinner and tails on the coin. Because they are independent events, it doesn't matter if we spin the spinner, then toss the coin, or toss the coin and then spin the spinner – or do them both at the same time.

The probability distributions for each are shown below:

Spinner   *Coin*
$P(\text{RED}) = \frac{1}{3}$ $P(\text{HEADS}) = \frac{1}{2}$
$P(\text{BLUE}) = \frac{1}{3}$ $P(\text{TAILS}) = \frac{1}{2}$
$P(\text{YELLOW}) = \frac{1}{3}$

When dealing with independent events we find the probability of both events happening simply by multiplying the two individual probabilities together.

$$P(\text{YELLOW, HEADS}) = \frac{1}{3} \times \frac{1}{2} = \frac{1}{6}$$

Worked Example 6

Three fair coins are tossed. What is the probability of getting exactly two heads?

Solution

Table 9.2 shows that there are 8 possible outcomes:

If S is the sample space, then $n(S) = 8$

There are 3 outcomes (outcomes 2, 3 and 5) in which there are exactly two heads.

If E is the event of getting exactly 2 heads, then $n(E) = 3$.

Therefore:

$$P(E) = \frac{n(E)}{n(S)} = \frac{3}{8}$$

Fig. 9.3 A 3-sided spinner and a coin

Table 9.2 The possible outcomes from throwing three coins

Coin 1	Coin 2	Coin 3	Number of heads
Heads	Heads	Heads	3
Heads	*Heads*	*Tails*	*2*
Heads	*Tails*	*Heads*	*2*
Heads	Tails	Tails	1
Tails	*Heads*	*Heads*	*2*
Tails	Heads	Tails	1
Tails	Tails	Heads	1
Tails	Tails	Tails	0

9.9 Random Variables

A **random variable** is a set of possible values from a random experiment. When the values are confined to discrete values, we refer to the random variable as a **discrete random variable**.

A random variable is not quite the same as the sample space, because we are able to *assign* a value to a particular outcome. A random variable provides a useful way of expressing a probability distribution.

Example
We saw that a 5-sided spinner can have five following possible outcomes: The spinner can land on 1, 2, 3, 4, or 5.

We know that the sample space for the spinner is: $S = \{1, 2, 3, 4, 5\}$.

In this case, as the outcomes are numbers, it makes sense to define our random variable X as:

$$X = \{1, 2, 3, 4, 5\}$$

We write the probability for each outcome as $P(X = x)$.
In the case of the spinner we can write:

$$P(X = 1) = \frac{1}{5} \quad P(X = 2) = \frac{1}{5} \quad P(X = 3) = \frac{1}{5}$$

$$P(X = 4) = \frac{1}{5} \quad P(X = 5) = \frac{1}{5}$$

The sum of the probabilities is, of course, 1.

Now consider the experiment that we looked at in the last worked example, namely tossing three coins. We are interested in the number of heads thrown, which is shown in Table 9.3.

9.10 Expected Value (Mean Value)

Table 9.3 Number of heads thrown when tossing three coins

Coin 1	Coin 2	Coin 3	Number of heads
Heads	Heads	Heads	3
Heads	Heads	Tails	2
Heads	Tails	Heads	2
Heads	Tails	Tails	1
Tails	Heads	Heads	2
Tails	Heads	Tails	1
Tails	Tails	Heads	1
Tails	Tails	Tails	0

Table 9.4 The probability distribution for number of heads thrown when tossing three coins

No of heads thrown, x	0	1	2	3
Number of occurrences	1	3	3	1
$P(X = x)$	1/8	3/8	3/8	1/8

The number of heads thrown can be 0, 1, 2, or 3. So the random variable representing the number of heads thrown is:

$$X = \{0, 1, 2, 3\}$$

This enables us to express the probability distribution as shown in Table 9.4. Consider the following events:

1. The event of throwing *exactly* two heads:

$$\text{Here we have: } P(X = 2) = \frac{3}{8}$$

2. The event of throwing *at least* two heads:

$$\text{In this case } P(X \geq 2) = \frac{3}{8} + \frac{1}{8} = \frac{1}{2}$$

9.10 Expected Value (Mean Value)

An **expected value** refers to the average value of the result of performing an experiment a very large number of times. We express the expected value as $E(x)$, where x is the value of a random variable defined for our experiment.

Table 9.5 The probability distribution for a fair 5-sided spinner

Number thrown, x	1	2	3	4	5
$P(X = x)$	1/5	1/5	1/5	1/5	1/5

Table 9.6 The probability distribution for a weighted dice

Number thrown, x	1	2	3	4	5	6
$P(X = x)$	0.3	0.2	0.2	0.1	0.1	0.1

Rather than actually throwing a dice, say, ten thousand times we can use a formula to do this calculation. The calculation involves taking each value (x), multiplying it by its probability (p) and adding them all together.

This is expressed as: $E(x) = \Sigma xp$.

Example
Table 9.5 represents the probability distribution for a fair 5-sided spinner:

$$E(x) = \frac{1}{5} \times 1 + \frac{1}{5} \times 2 + \frac{1}{5} \times 3 + \frac{1}{5} \times 4 + \frac{1}{5} \times 5 = 3$$

For a fair spinner, this is exactly the result we would expect to find.

Now let's consider the probability for a dice which is weighted as follows (Table 9.6).

$$E(x) = 0.3 \times 1 + 0.2 \times 2 + 0.2 \times 3 + 0.1 \times 4 + 0.1 \times 5 + 0.1 \times 6 = 2.8$$

In this case we see that the expected value is not in our sample set—this is to be expected, because it is an average.

Worked Example 7
Consider the previous example of tossing a coin three times, where X was the random variable representing the number of heads thrown. What is the expected value?

Solution
We saw that the probability distribution was as shown in Table 9.4. The expected value is calculated as follows:

$$E(x) = \frac{1}{8} \times 1 + \frac{3}{8} \times 3 + \frac{3}{8} \times 3 + \frac{1}{8} = 1.5$$

9.11 Conditional Probability

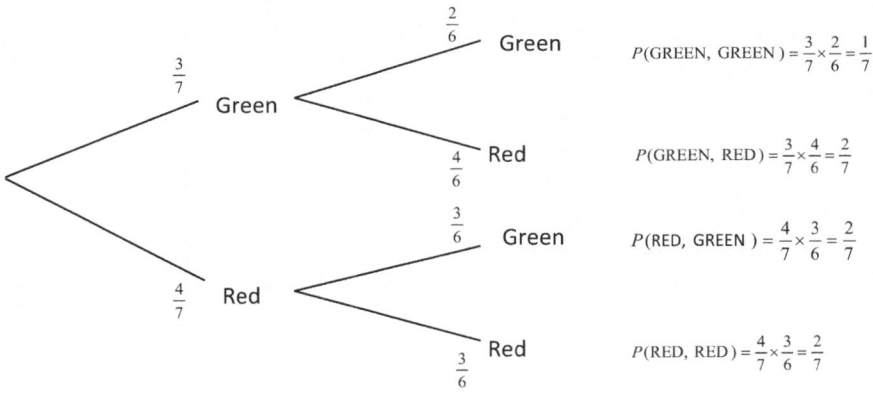

Fig. 9.4 A tree diagram

9.11 Conditional Probability

Sometimes the outcome of an event is dependent on another event. As an example, imagine that a bag contains 3 green balls and 4 red balls. What is the probability of picking a green ball, followed by a red ball?

The probability of picking a green ball is $\frac{3}{7}$.

If we have picked a green ball, then there are 2 green balls left and 4 red balls (6 in all). So the probability of picking a red ball is now $\frac{4}{6}$.

Therefore, the probability of picking a green ball followed by a red ball is:

$$\frac{3}{7} \times \frac{4}{6} = \frac{2}{7}$$

9.11.1 Tree Diagrams

Tree diagrams provide a useful way of representing conditional probability. The diagram shown in Fig. 9.4 represents the previous example of randomly selecting balls from a bag that contains three green balls and four red balls, as in the last example. It can be seen that the diagram shows the probability of all possible events, obtained by multiplying the individual probabilities.

Worked Example 8
A bag contains three green balls, three yellow balls and four red balls.

(a) A ball is picked randomly from the bag, and is then returned to the bag. Another ball is then picked. What is the probability of picking a green ball, followed by a red ball?
(b) A ball is picked randomly from the bag, but is not returned. Another ball is then picked. What is the probability of picking a green ball, followed by another green ball?
(c) Represent the probabilities described in part (b) on a tree diagram.
(d) Use this diagram to find the probability of picking a red ball followed by a yellow ball OR a yellow ball followed by a green ball.

Solution

(a) In this case the events are independent.
The probability of picking a green ball is:
$$\frac{3}{10}$$
The probability of picking a red ball is:
$$\frac{4}{10}$$
The probability of picking a green ball followed by a red ball is:
$$\frac{3}{10} \times \frac{4}{10} = \frac{3}{25}$$

(b) In this case, the events are not independent.
The probability of picking a green ball is:
$$\frac{3}{10}$$
There are 9 balls left, one green ball has gone, so the probability of another green ball is:
$$\frac{2}{9}$$
So the probability of picking a green ball followed by another green ball is:
$$\frac{3}{10} \times \frac{2}{9} = \frac{1}{15}$$

(c) The tree diagram is shown in Fig. 9.5

9.12 Bayes' Theorem

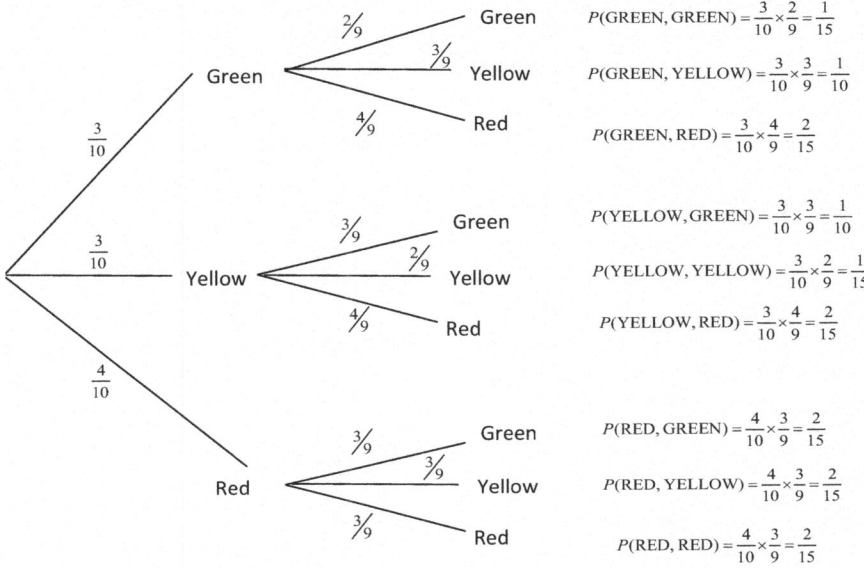

Fig. 9.5 Solution to worked example 8(c)

(d) The probability of picking a red ball followed by a yellow ball OR a yellow ball followed by a green ball is:

$$\frac{2}{15} + \frac{1}{10} = \frac{7}{30}$$

9.12 Bayes' Theorem

Bayes' theorem (named after Thomas Bayes) is an extremely important theorem in probability and statistics. It is used extensively in such areas as finance, epidemiology and drug testing. It is based on conditional probability, and it is used to determine the probability of an event based on information about other events.

Bayes' Theorem can be stated as follows:

$$P(A|B) = \frac{P(A) \times P(B|A)}{P(B)}$$

Here $P(A|B)$ is read as the probability of A given B, and refers to the probability that event A will happen if we already know that event B has happened.

We will not show you the proof of Bayes' theorem here, but this can easily be looked up for those who are interested.

To better understand Bayes' theorem, let us return to the example given in Sect. 9.11. A bag contains 3 green balls and 4 red balls.

The probability of picking a green ball is:

$$P(\text{GREEN}) = \frac{3}{7}$$

Similarly, the probability of picking a red ball is

$$P(\text{RED}) = \frac{4}{7}$$

If a red ball has already been picked then the probability of picking a green ball becomes $\frac{3}{6}$ or $\frac{1}{2}$. We write this as:

$$P(\text{GREEN}|\text{RED}) = \frac{1}{2}$$

This is read as the probability of GREEN given RED.

Bayes' theorem tells us:

$$P(\text{RED}|\text{GREEN}) = \frac{P(\text{RED}) \times P(\text{GREEN}|\text{RED})}{P(\text{GREEN})}$$

$$= \frac{\frac{4}{7} \times \frac{1}{2}}{\frac{3}{7}}$$

$$= \frac{4}{7} \times \frac{1}{2} \times \frac{7}{3} = \frac{2}{3}$$

In this case, all the information is known to us, so we could have worked out for ourselves the probability of choosing a red ball once a green ball has been removed—in other words $P(\text{RED}|\text{GREEN})$; and it is indeed $\frac{4}{6}$ or $\frac{2}{3}$, thus showing that Bayes' theorem holds in this particular case.

There are, however, many cases where we do not have all the information, and this is when Bayes' theorem comes into its own.

As an example, consider the following:

- a test for a disease has a 95% probability of yielding a correct positive result when the disease is present;
- 2% of all people who are tested test positive.
- it is known that 1.5% of the population actually has the disease.

We need to know the probability that an individual who tests positive actually has the disease. In other words we need to calculate $P(\text{DISEASE}|\text{POSITIVE})$.

Bayes' theorem tells us that:

$$P(\text{DISEASE}|\text{POSITIVE}) = \frac{P(\text{DISEASE}) \times P(\text{POSITIVE}|\text{DISEASE})}{P(\text{POSITIVE})}$$

9.12 Bayes' Theorem

From the information given, we know that:
The probability of someone having the disease is:

$$P(\text{DISEASE}) = 0.015$$

The probability of someone testing positive when they have the disease is:

$$P(\text{POSITIVE}|\text{DISEASE}) = 0.95$$

The probability of anyone testing positive is:

$$P(\text{POSITIVE}) = 0.02$$

Putting these into our formula gives us:

$$P(\text{DISEASE}|\text{POSITIVE}) = \frac{0.015 \times 0.95}{0.02}$$
$$\approx 71\%$$

Worked Example 9
A college has 100 students. 40 are men and 60 are women. 20 of the men study science subjects, and 50 of the women study science subjects.

(a) What is the probability of randomly selecting a man from the student body?
(b) What is the probability of randomly selecting a student who studies science?
(c) What is the probability of randomly selecting a student who studies science from the male students?
(d) What is the probability of randomly selecting a male student from those who study science?
(e) Show that Bayes' theorem holds for this set of figures.

Solution

(a) $P(\text{MAN}) = \frac{40}{100} = \frac{2}{5}$.
(b) $P(\text{SCIENCE}) = \frac{70}{100} = \frac{7}{10}$.
(c) $P(\text{SCIENCE}|\text{MAN}) = \frac{20}{40} = \frac{1}{2}$.
(d) $P(\text{MAN}|\text{SCIENCE}) = \frac{20}{70} = \frac{2}{7}$.
(e) Bayes' theorem tells us:

$$P(\text{SCIENCE}|\text{MAN}) = \frac{P(\text{SCIENCE}) \times P(\text{MAN}|\text{SCIENCE})}{P(\text{MAN})}$$
$$= \frac{\frac{7}{10} \times \frac{2}{7}}{\frac{2}{5}}$$

$$= \frac{1}{2}$$

This is the same result as we got for part (c), showing that Bayes' theorem holds.

Worked Example 10

A study finds that the probability of a particular word appearing in an email is 15%. It also finds that the probability of an email being spam if it contains this word is 40%. The probability of any email being spam is 20%.

Use Bayes' theorem to find the probability that a spam email contains this particular word.

Solution

$$P(\text{WORD}) = 0.15$$
$$P(\text{SPAM}) = 0.2$$
$$P(\text{SPAM}|\text{WORD}) = 0.4$$

$$P(\text{WORD}|\text{SPAM}) = \frac{P(\text{WORD}) \times P(\text{SPAM}|\text{WORD})}{P(\text{SPAM})}$$
$$= \frac{0.15 \times 0.4}{0.2}$$
$$= 30\%$$

9.13 Binomial Probability

Imagine an experiment with exactly two mutually exclusive outcomes, such as flipping a coin, and imagine we repeat this experiment several times. Each experiment is called a Bernoulli trial (after the Swiss mathematician, Jacob Bernoulli). In a series of Bernoulli trials, the repeated actions must all be independent. In fact, we don't have to restrict the experiment to ones with only two "natural" outcomes, because when there are multiple outcomes, we can define one outcome as *success* and all others as *failure*.

So the following are all examples of Bernoulli trials:

- throwing a dice, where getting a 6 is a success, all other outcomes are failures;
- asking a question in an opinion poll, where "yes" is considered success, but "no", "don't know" and "won't vote" are failures;
- answering a question in a multiple choice exam, where one answer is correct (success) all other answers are incorrect (failure).

9.13.1 The Binomial Probability Formula

There is a formula we can use to calculate the number of successful outcomes in a series of Bernoulli trials:

$$P(k \text{ successes in } n \text{ trials}) = {}^nC_k p^k q^{n-k}$$

n	total number of trials
k	number of successes
$n - k$	number of failures
p	probability of success in one trial
$q = 1 - p$	probability of failure in one trial

The examples that follow will make this clear.

Worked Example 11
A multiple choice paper consists of 10 questions, each with five possible answers. Only one answer is correct.

If the questions are answered completely randomly, what is the probability of getting *exactly* four correct answers?

Solution
We use the formula:

$$P(k \text{ successes in } n \text{ trials}) = {}^nC_k p^k q^{n-k}$$

In this case:

$$n = 10$$
$$k = 4$$
$$n - k = 6$$
$$p = 0.2 \text{ (There are 5 correct answers)}$$
$$q = 0.8$$

$$P(k \text{ successes in } n \text{ trials}) = {}^nC_k p^k q^{n-k}$$

$$\begin{aligned} P(4 \text{ successes in } 10 \text{ trials}) &= {}^{10}C_4 \times 0.2^4 \times 0.8^6 \\ &= 210 \times 0.0016 \times 0.262144 \\ &= 0.088 \end{aligned}$$

Note that if you had been asked the probability of getting *at least* four correct answers, you would have to add all the probabilities from four correct answers up to 10 correct answers.

Worked Example 12
A coin is tossed five times. What is the probability of getting exactly three heads if:

(a) the coin is a fair coin;
(b) the coin is weighted so that it is three times more likely to land on tails than on heads?

Solution
In both cases we use the formula:

$$P(k \text{ successes in } n \text{ trials}) = {}^nC_k p^k q^{n-k}$$

In each case:

$$n = 5$$
$$k = 3$$
$$n - k = 2$$

(a) In this case:

$$p = 0.5$$
$$q = 0.5$$

$$\begin{aligned} P(3 \text{ successes in 5 trials}) &= {}^5C_3 \times 0.5^3 \times 0.5^2 \\ &= 10 \times 0.125 \times 0.25 \\ &= 0.3125 \end{aligned}$$

(b) In this case:

$$p = 0.5$$
$$q = 0.75$$

$$P(3 \text{ successes in } 5 \text{ trials}) = {}^5C_3 \times 0.25^3 \times 0.75^2$$
$$= 10 \times 0.15625 \times 0.5625$$
$$= 0.00879$$

9.14 Application to Computing

There are many important areas of computer science in which probability plays a crucial role. It is used in algorithms for machine learning to model uncertainty and to make predictions. It is also plays an important role in data analysis. In cryptography it is used to analyze the security of protocols and algorithms. In network security probability is used to assess the likelihood of cyber attacks and to design secure communication protocols.

9.15 Exercises

1. A 5-sided spinner has sides coloured yellow, blue, green, red and black.
    (a) What is the sample space, S?
    (b) Give the value of E, the event that the spinner lands on a colour which starts with the letter "B".
    (c) What is the cardinality of the set S and the set E?
    (d) Calculate the probability that E happens.
2. A small lottery consists of randomly choosing one of twelve balls, labelled 1–12.
    The following events are defined:
    A is the event of choosing a number less then 8.
    B is the event of choosing an odd number.
    C is the event of choosing a number divisible by 4.
    (a) What is the sample space, S?
    (b) Give the value of A, B and C.
    (c) Give the values of: $A \cap B$ $A \cap C$ $B \cap C$
    (d) Which pairs of events are mutually exclusive?
    (e) Calculate the probability of choosing an odd number or a number divisible by 4.
    (f) Calculate the probability of choosing a number less than 8 or a number divisible by 4.
3. An experiment is done, and two events, A and B are defined.
    The probability of A happening is 0.6.
    The probability of B happening is 0.4.
    The probability of both A and B happening is 0.2.
    What is the probability of A or B happening?
4. Consider the following reduced deck of cards consisting of just the following:

2♣, 3♣, Q♣, A♣
4♦, 10♦, Q♦
2♥, 7♥, Q♥, A♥
2♠, 5♠, K♠, A♠

What is the probability of drawing a card that is black or is an ace?
5. An unfair 5-sided spinner with colours red, yellow, blue, green and orange is weighted as follows: The chance of the spinner landing on green is twice that of it landing on a orange. The chance of it landing on blue is 3 times that of landing on orange. It is 4 times more likely to land on yellow than it is to land on a orange. It is 10 times more likely to land on red than to land on orange.
   (a) Find the probability distribution of the spinner.
   (b) What is the probability of the spinner landing on blue or green?
6. Two 3-sided spinners, each with sides coloured red, blue and yellow are spun. What is the probability of at least one of the spinners landing on yellow?
7 Two 3-sided spinners are spun, each with colours yellow, red and blue. If they are fair spinners, what is the expected value of getting at least one yellow?
8. A bag contains 4 blue balls, 5 yellow balls and 6 red balls.
   (a) A ball is picked randomly from the bag, and is then returned to the bag. Another ball is then picked. What is the probability of picking a blue ball, followed by a red ball?
   (b) A ball is picked randomly from the bag, but is not returned. Another ball is then picked. What is the probability of picking a red ball, followed by another red ball?
      (i) Represent the probabilities described in part b) on a tree diagram.
      (ii) Use this diagram to find the probability of picking a red ball followed by a yellow ball OR a yellow ball followed by a blue ball.
9. A bag contains a red ball and a white ball. A second bag contains a red ball, a white ball and a yellow ball. A ball is picked randomly from each bag.
   (a) What is the sample space, S?
   (b) Give the value of E, the event that a red ball and a white ball are picked.
   (c) Give the value of F, the event that a red ball and a yellow ball are picked.
   (d) What is the cardinality of the sets S, E and F?
   (e) Calculate the probability that E happens.
   (f) Calculate the probability that F happens.
10 Polling in a constituency showed that 50% of the electorate intended to vote for one particular party. Of those, 35% were under the age of 40. If the percentage of people under 40 who are of voting age in that constituency is 30%, use Bayes' theorem to find the probability that a person who votes for that party is under 40 years old.
11 A multiple choice paper consists of 12 questions, each with 4 possible

answers. Only one answer is correct.

If the questions are answered completely randomly, what is the probability of getting *exactly* 50% (6 correct answers)?

12. A dice is thrown 4 times. What is the probability of getting exactly 3 sixes if
    (a) the dice is a fair dice;
    (b) the dice is weighted so that it is three times more likely to land on a 5 or a 6 than on any other number?

Graph Theory   10

At the end of this chapter you should be able to:

- explain what is meant by the terms **graph**, **vertex**, **node**, **edge** and **adjacency matrix**;
- distinguish between a **simple graph** and a **multigraph**, and between a **connected graph** and an **unconnected graph**;
- explain what is meant by the **degree** of a vertex, and state the sum of **degrees of vertices theorem**;
- explain what is meant by the **distance** between two vertices, the **eccentricity** of a vertex and the **radius** and **diameter** of a vertex;
- define the term **subgraph**;
- distinguish between **paths**, **trails**, **circuits** and **cycles**;
- identify **isomorphic** and **homeomorphic** graphs;
- define the terms **traversable trail** and **traversable graph** and explain the terms **Eulerian** graph and **Hamiltonian** graph;
- describe what is meant by a **weighted** graph and provide an adjacency matrix for a weighted graph;
- explain what is meant by a **tree** and a **binary** tree;
- explain the term **spanning** tree, and find a **minimum spanning tree** for a particular graph using **Kriskal's algorithm**;
- describe three methods for traversing a binary tree;
- explain the term **planar** graph, and state and use **Euler's formula**;
- describe what is meant by a **directed graph (digraph)**, and provide adjacency matrices for directed graphs.

10.1 Introduction

In this final chapter we will be studying graph theory. Here the word graph does not refer to the most common usage of the word—namely a chart showing the relationship between two or more quantities by means of a curve or line—but to a representation of connected objects. As you will see at the end of the chapter, graph theory plays a very important role in the world of computer science.

10.2 Definitions

A **graph** is a pictorial representation of a set of connected objects. An example is shown in Fig. 10.1.

The connected objects are represented by points. These points are known as **vertices** or **nodes**. The links that connect the vertices are called **edges**.

It is usual to name the vertices using lower case letters. In Fig. 10.1, the vertices are a, b, c and d. The edges are e_1, e_2, e_3 and e_4.

Formally, we can say that a graph, G, consists of a set of vertices, V, and a set of edges, E:

$$G = (V, E)$$

In Fig. 10.1:

$$V = \{a, b, c, d\} \quad E = \{e_1, e_2, e_3, e_4\}$$

Vertices such as a and b in Fig. 10.1 which are connected by an edge are said to be **adjacent**. A graph can be represented by an **adjacency matrix**, in which each element of the matrix is either a 0, showing that there is no direct connection

Fig. 10.1 A simple graph

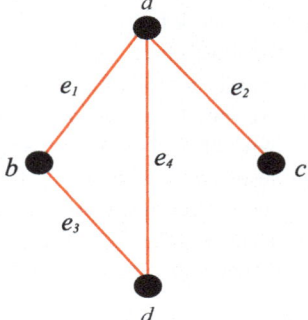

10.4 Connected and Unconnected Graphs

Fig. 10.2 A multigraph

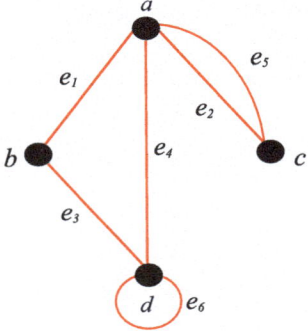

between the vertices, or 1 showing that there is a connection. The adjacency matrix below represents the graph shown in Fig. 10.1.

$$\begin{array}{c} \\ a \\ b \\ c \\ d \end{array} \begin{array}{c} a\ b\ c\ d \\ \begin{pmatrix} 0 & 1 & 1 & 1 \\ 1 & 0 & 0 & 1 \\ 1 & 0 & 0 & 0 \\ 1 & 1 & 0 & 0 \end{pmatrix} \end{array}$$

10.3 Multigraphs

In the graph shown in Fig. 10.1, each pair of vertices is connected by one edge only. Now consider the graph shown in Fig. 10.2.

In Fig. 10.2, the vertices a and c are connected by two edges, e_2 and e_5. These two edges are referred to as **multiple** edges. Additionally one edge, e_6, has the same start point and endpoint—such an edge is referred to as a **loop**. A graph that allows multiple edges and loops is called a **multigraph**.

10.4 Connected and Unconnected Graphs

The graphs shown in Figs. 10.1 and 10.2 are **connected** graphs, because there is a path between any two of the vertices.

However, a graph such as that shown in Fig. 10.3 is an **unconnected** graph, because, for example, there is no path between vertex a and vertex d.

Fig. 10.3 An unconnected graph

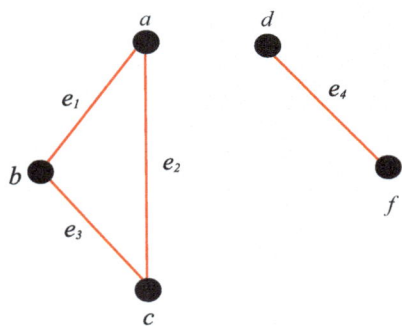

10.5 Degree of a Vertex

The degree of a vertex is the number of vertices that are adjacent to that vertex. The degree of a vertex, v, for example, is written:

$$deg(v)$$

In Fig. 10.1 we have:

$$deg(a) = 3 \quad deg(b) = 2 \quad deg(c) = 2 \quad deg(d) = 2$$

In a simple graph with n vertices, the degree of any vertex is always less than or equal to $n - 1$.

10.5.1 Sum of Degrees of Vertices Theorem

The sum of the degrees of the vertices is equal to twice the number of edges. So in Fig. 10.1, for example, the sum of the degrees of vertices is 8, which is twice the number of edges, 4.

This also holds for multigraphs, but we must remember to count the degree of a vertex which has a loop as 2.

So in Fig. 10.2 we have:

$$deg(a) = 4 \quad deg(b) = 2 \quad deg(c) = 2 \quad deg(d) = 4$$

Thus the sum of the degrees is 12, which is twice the number of edges, 6.

10.6 Distance Between Two Vertices

The distance between two vertices u and v is the length of the shortest path between u and v. It is written $d(u,v)$.

Fig. 10.4 A graph with 6 vertices and 7 edges

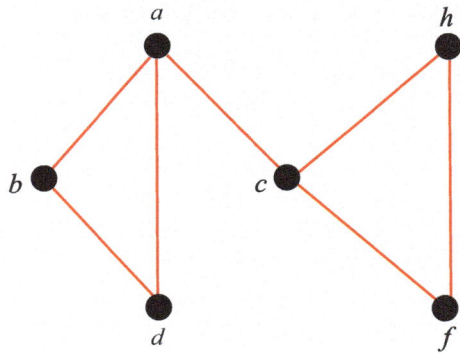

For example, in Fig. 10.4 we have:

$$d(a,f) = 2 \quad d(h,f) = 1 \quad d(b,f) = 3.$$

10.7 Eccentricity of a Vertex

The eccentricity of a vertex, v, is the highest distance from a particular vertex to any other vertex. It is written $e(v)$.

For example, in Fig. 10.5:

$$e(b) = 4 \ (b \text{ to } a, a \text{ to } c, c \text{ to } f, f \text{ to } g)$$
$$e(a) = 3 \ (a \text{ to } c, c \text{ to } f, f \text{ to } g)$$
$$e(c) = 2 \ (c \text{ to } f, f \text{ to } g \text{ or } c \text{ to } a, a \text{ to } b).$$

Fig. 10.5 A graph with 7 vertices and 6 edges

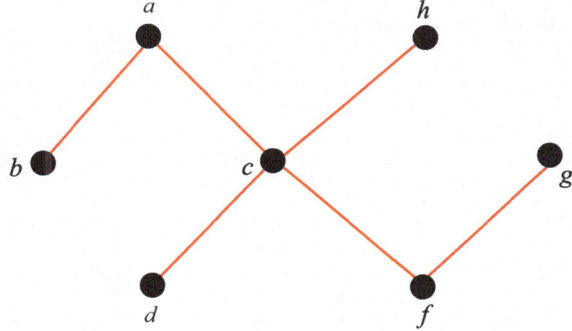

10.8 Radius and Diameter

The **radius** of a graph G is the lowest eccentricity of that graph.
For example, in Fig. 10.5:

$$radius(G) = 2.$$

The **diameter** of a graph G is the highest eccentricity of that graph.
For example, in Fig. 10.5:

$$diameter(G) = 4.$$

Worked Example 1
Consider the graph shown in Fig. 10.6:

(a) State the value of:

$$deg(a) \ deg(b) \ deg(c) \ deg(d) \ deg(e) \ deg(f) \ deg(g)$$

(b) Show that the *sum of degrees of vertices* theorem holds for this graph.
(c) State the value of:

$$d(a,g) \quad d(a,d) \quad d(b,g)$$

(d) State the value of:

$$e(g) \quad e(a) \quad e(c)$$

(e) Referring to the graph in Fig. 10.6 as G, give the values of:

$$radius(G) \quad diameter(G).$$

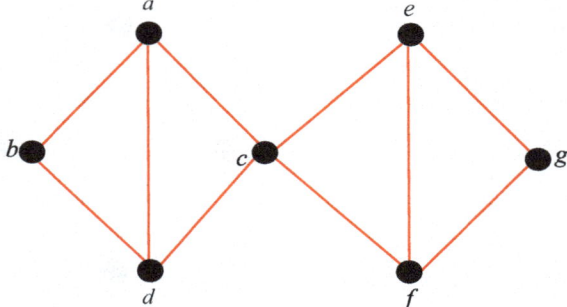

Fig. 10.6 Worked example 1

10.10 Paths, Trails, Circuits and Cycles

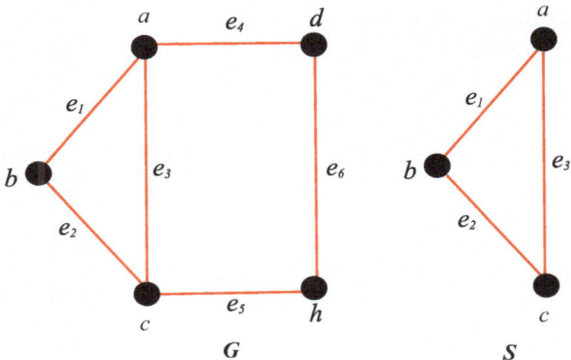

Fig. 10.7 A graph and a subgraph

Solution

(a) $deg(a) = 3 \quad deg(b) = 2 \quad deg(c) = 4 \quad deg(d) = 3$
$deg(e) = 3 \quad deg(f) = 3 \quad deg(g) = 2$
(b) Sum of the degrees of vertices is 20, which is twice the number of edges, 10.
(c) $d(a, g) = 3 \quad d(a, d) = 1 \quad d(b, g) = 4$
(d) $e(g) = 4 \quad e(a) = 3 \quad e(c) = 2$
(e) $radius(G) = 2 \quad diameter(G) = 4$.

10.9 Subgraphs

Figure 10.7 shows a graph G and a graph S. S is a **subgraph** of G—the vertices and edges of S are contained in the vertices and edges of G.

Formally:

if $S = (V_2, E_2)$ and $G = (V_1, E_1)$ then S is a subgraph of G if:

$$V_2 \subset V_1 \quad \text{and} \quad E_2 \subset E_1$$

In Fig. 10.7:

$V_2 = \{a, b, c\}$ and $V_1 = \{a, b, c, d, h\}$ so $V_2 \subset V_1$
$E_2 = \{e_1, e_2, e_3\}$ and $E_1 = \{e_1, e_2, e_3, e_4, e_5, e_6\}$ so $E_2 \subset E_1$.

10.10 Paths, Trails, Circuits and Cycles

A **path** in a graph is a sequence of alternating vertices and edges. Consider the graph shown in Fig. 10.8.

Fig. 10.8 A graph with 7 vertices and 8 edges

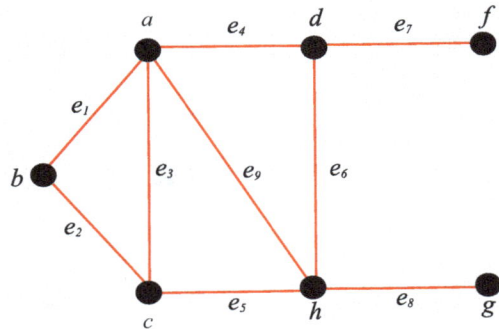

The following are all possible paths in this graph—we have assigned each a number in order to refer to them later:

1: $a, e_3, c, e_5, h, e_8, g$
2: a, e_4, d, e_7, f
3: $c, e_2, b, e_1, a, e_3, c, e_5, h$
4: $c, e_2, b, e_1, a, e_3, c, e_2, b$
5: $d, e_4, a, e_1, b, e_2, c, e_3, a, e_9, h, e_6, d$
6: $b, e_1, a, e_3, c, e_2, b$
7: $a, e_3, c, e_5, h, e_6, d$
8: a, e_4, d

The number of edges in a path is called the **length** of the path. For example, path 3 above has a length of 4, while path 8 has a length of 1.

A path in which all vertices are distinct is called a **simple path**. Paths 1, 2, 7 an 8 above are all simple paths.

A path in which all edges are distinct is called a **trail**. All of the above paths are trails, with the exception of path 4.

A path in which all edges are distinct (that is, a trail), and which begins and ends on the same vertex is called a **circuit**. Paths 5 and 6 above are circuits.

A circuit in which all vertices (other than the start and end vertex) are distinct is called a **cycle**. Path 6 above is a cycle, but path 5 is not. A cycle with n edges is referred to as an n-cycle. For example path 5 above is a 3-cycle.

Paths 7 and 8 are both paths from a to d. However, 8 is the *shortest* path from a to d.

Worked Example 2
Consider the graph shown in Fig. 10.9:
 Now consider the following path: $f, e_6, a, e_5, d, e_3, b, e_1, a, e_6, f$.
 State whether this path is:

(a) a simple path;

10.11 Isomorphic and Homeomorphic Graphs

Fig. 10.9 Worked example 2

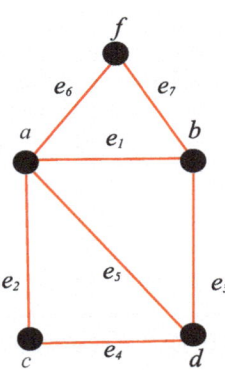

(b) a trail;
(c) a circuit;
(d) a cycle.

Solution

(a) It is not a simple path because the vertices are not all distinct.
(b) It is not a trail, because not all edges are distinct.
(c) It is not a circuit, because not all edges are distinct.
(d) It is not a cycle, because a cycle is a type of circuit—also not all vertices are distinct.

10.11 Isomorphic and Homeomorphic Graphs

A graph G_1 and a graph G_2 are said to be **isomorphic** if every edge of G_1 is an edge of G_2, and vice versa. In other words, they are exactly the same graph, but are drawn differently. Two examples are given in Fig. 10.10.

If two graphs can be obtained from the same graph by dividing an edge with additional vertices, they are said to be **homeomorphic**. For example, in Fig. 10.11 graphs (b) and (c) are homeomorphic because they can both be obtained from graph (a).

Fig. 10.10 Isomorphic graphs

Fig. 10.11 Graphs (*b*) and (*c*) are homoeomorphic because they can both be obtained from graph (*a*)

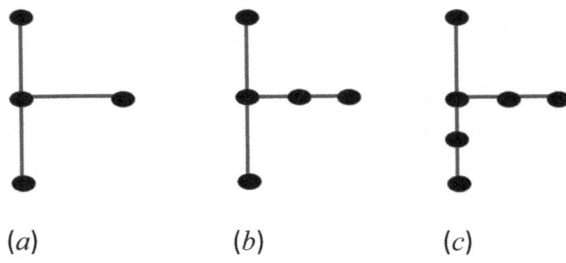

(*a*)   (*b*)   (*c*)

10.12 Traversable Graphs

A graph is said to be **traversable** if it is possible to draw a path that includes all of the vertices and all of the edges, without repeating any of the edges. Such a path is by definition a trail (since all edges are distinct) and is referred to as a **traversable trail**.

The graph shown in Fig. 10.12 is traversable—it contains, for example, the following traversable trail:

$$f, e_5, c, e_4, d, e_1, a, e_2, b, e_3, c$$

The following theorem holds:

A connected graph is traversable if and only if the number of vertices with odd degree is exactly 2 or 0.

In Fig. 10.12, vertices c and f are odd. By contrast, the graph shown in Fig. 10.13 has four odd vertices (c, d, f, g) and is not traversable.

Fig. 10.12 A traversable graph

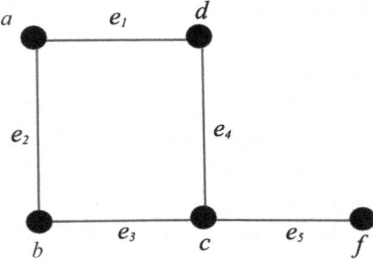

10.12 Traversable Graphs

Fig. 10.13 A non-traversable graph

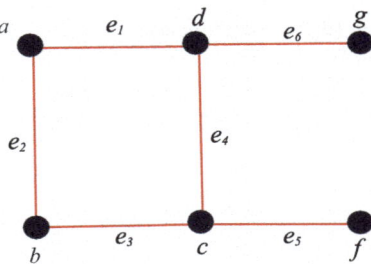

Fig. 10.14 An Eulerian graph

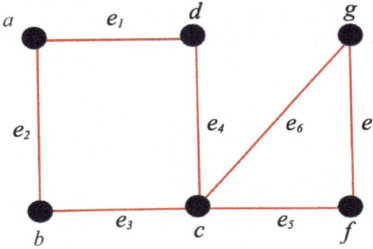

10.12.1 Eulerian Graphs

A traversable trail that begins and ends on the same vertex is called an **Eulerian**[1] **trail** or **Eulerian circuit**. A graph that contains an Eulerian circuit is called an **Eulerian graph**.

The graph shown in Fig. 10.14 is Eulerian: it contains the following Eulerian circuit:

$$a, e_1, d, e_4, c, e_6, g, e_7, f, e_5, c, e_3, b, e_2, a$$

The following holds:
A connected graph is Eulerian if and only if each vertex has an even degree.

10.12.2 Hamiltonian Graphs

We saw above that an Eulerian circuit traverses each edge exactly once but may repeat vertices.

In contrast a path that begins and ends with the same vertex and visits all the vertices in a graph exactly once (apart from the beginning and end vertex) is called a **Hamiltonian circuit**. A Hamiltonian circuit does not need to contain every edge. A graph that contains a Hamiltonian circuit is called a **Hamiltonian graph**.

---

[1] Pronounced "Oilerian".

Fig. 10.15 A Hamiltonian graph

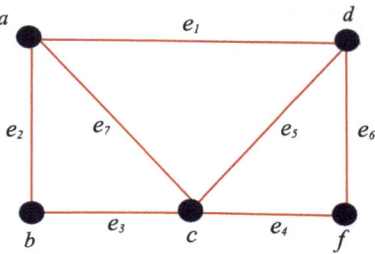

The graph shown in Fig. 10.15 is a Hamiltonian graph, because it contains, for example, the following Hamiltonian circuit:

$$b, e_3, c, e_4, f, e_6, d, e_1, a, e_2, b$$

Worked Example 3

State whether the graph shown in Fig. 10.16 is:

(a) Traversable;
(b) Eulerian
(c) Hamiltonian.

Solution

(a) It is traversable because exactly two of the vertices (c and d) have an odd degree. One traversable trail is:

$$c, e_2, a, e_6, f, e_7, b, e_1, a, e_8, d, e_4, c, e_5, b, e_3, d, e_4, c$$

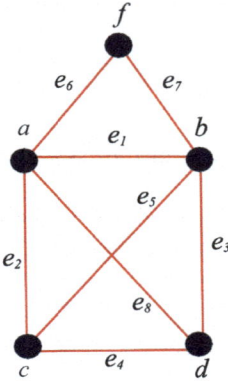

Fig. 10.16 Worked example 3

(b) It is not Eulerian because not all vertices have an even degree (c and d have odd degrees).
(c) It is Hamiltonian because it contains, for example, the following circuit, which visits all vertices:

$$c, e_4, d, e_3, b, e_7, f, e_6, a, e_2, c.$$

10.13 Weighted Graphs

A **weighted graph** is a graph in which each edge is associated with a numerical value. The weights can be anything relevant to a particular problem—for example, lengths, densities, costs, probabilities, durations and so on. The total of all the weights is referred to as the **weight** of the graph.

An example of a weighted graph is shown in Fig. 10.17.

A weighted graph can be represented by an adjacency matrix, as shown below. Each element of the matrix represents the weight of the edge connecting the two vertices.

$$\begin{array}{c} \\ a \\ b \\ c \\ d \\ e \end{array} \begin{array}{c} a\ b\ c\ d\ e \\ \begin{pmatrix} 0 & 2 & 0 & 3 & 0 \\ 2 & 0 & 5 & 0 & 0 \\ 0 & 5 & 0 & 6 & 0 \\ 3 & 0 & 6 & 0 & 1 \\ 0 & 0 & 0 & 1 & 0 \end{pmatrix} \end{array}$$

Worked Example 4
Provide an adjacency matrix for the weighted graph shown in Fig. 10.18.

Fig. 10.17 A weighted graph

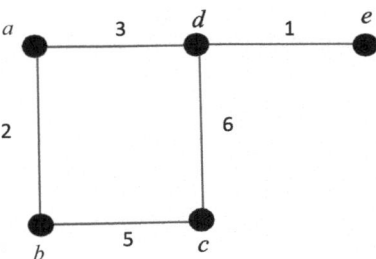

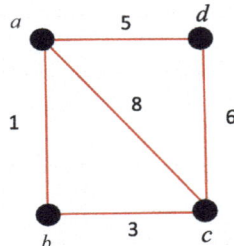

Fig. 10.18 Worked example 4

Solution

$$\begin{array}{c} \,a\,b\,c\,d \\ \begin{array}{c}a\\b\\c\\d\end{array}\begin{pmatrix} 0 & 1 & 8 & 5 \\ 1 & 0 & 3 & 0 \\ 8 & 3 & 0 & c \\ 5 & 0 & 6 & 0 \end{pmatrix}. \end{array}$$

10.14 Trees

A *connected acyclic graph* is called a **tree**. Put more simply, a tree is a connected graph with no cycles. An example of a tree is shown in Fig. 10.19.

The edges of a tree are known as **branches**. A node such as a is said to have two **child** nodes; a is referred to as the **parent** of b and d. The nodes without child nodes (for example b, c and g in Fig. 10.19) are called **leaf nodes**.

A tree with n vertices will have $n-1$ edges.

A *disconnected acyclic graph*—that is to say a disjoint collection of trees—is called a **forest**. Figure 10.20 shows an example of a forest.

Fig. 10.19 A tree

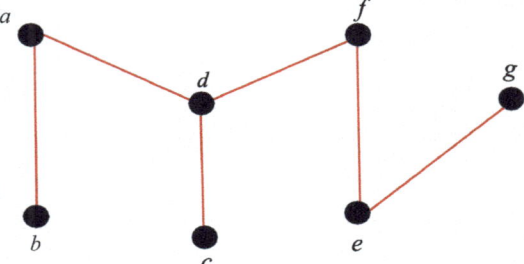

10.14 Trees

Fig. 10.20 A forest

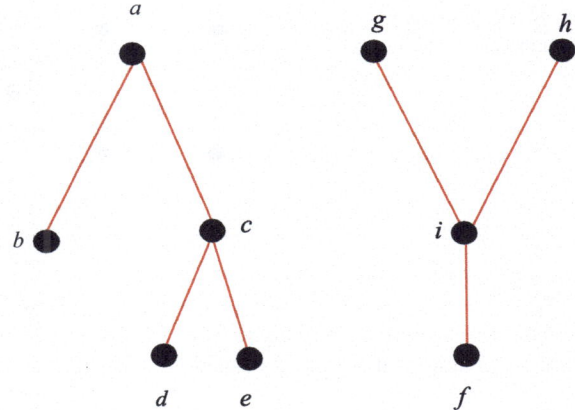

10.14.1 Spanning Trees

If a graph H is a subgraph of a graph G, then H is a **spanning tree** of G if H is a tree, and it contains all the vertices of G. In Fig. 10.21, graphs H and K are both spanning trees of graph G (there are others).

10.14.1.1 Minimum Spanning Tree

In a weighted graph, the minimum spanning tree is a spanning tree whose total weight is as small as possible. Figure 10.22 shows a weighted graph with its minimum spanning tree, which in this case has a weight of 12.

You should note that there can be more than one minimum spanning tree.

One way to find a minimum spanning tree is to apply **Kriskal's algorithm** which involves the following steps:

1. Sort all the edges from low weight to high.
2. Take the edge with the lowest weight and add it to the spanning tree.
3. Keep adding edges in order, but do not add an edge if it creates a cycle.
4. Keep adding edges in this way until all the vertices have been reached.

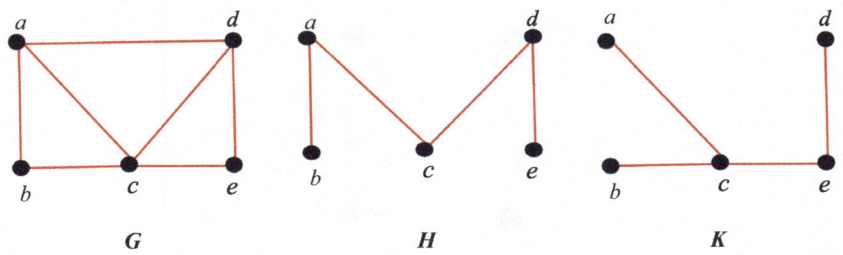

Fig. 10.21 Spanning trees

Fig. 10.22 Minimum spanning tree

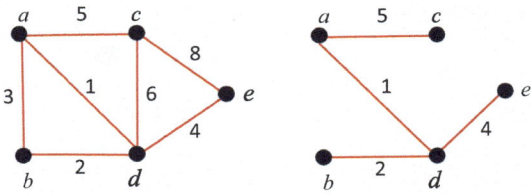

Let's apply this to the graph in Fig. 10.22.
We arrange the edges in order of weight, as shown in Table 10.1:
Add edges in order of weight (Fig. 10.23):
We can stop at this point because we have reached all the vertices.

Worked Example 5
Use Kriskal's algorithm to find the minimum spanning tree for the weighted graph shown in Fig. 10.24.

Solution
Arrange the edges in order of weight as in Table 10.2.

Table 10.1 First stage in Kriskal's algorithm

ad	bd	ab	de	ac	cd	ce
1	2	3	4	5	6	8

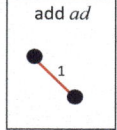
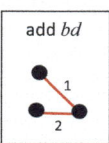

reject *ab* because it would create a cycle

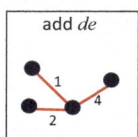
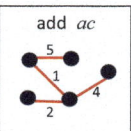

Fig. 10.23 Applying Kriskal's algorithm

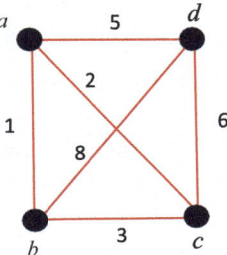

Fig. 10.24 Worked example 5

10.14 Trees

Table 10.2 Applying Kriskal's algorithm—worked example 5

ab	ac	bc	ad	cd	bd
1	2	3	5	6	8

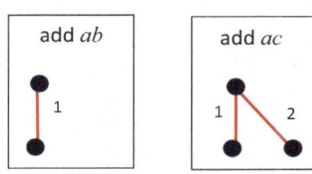
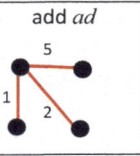

Fig. 10.25 Solution to worked example 5

Add edges in order of weight as shown in Fig. 10.25.
We can finish here because we have reached all the vertices.

10.14.2 Binary Trees

A **binary tree** is a tree which has a root node that has exactly two children, and in which each of the subsequent nodes has at most two children. Binary trees are very important data structures in computer science, and organizing data in this way provides very useful algorithms for such things as searching and sorting. Figure 10.26 shows an example of a binary tree.

Fig. 10.26 A binary tree

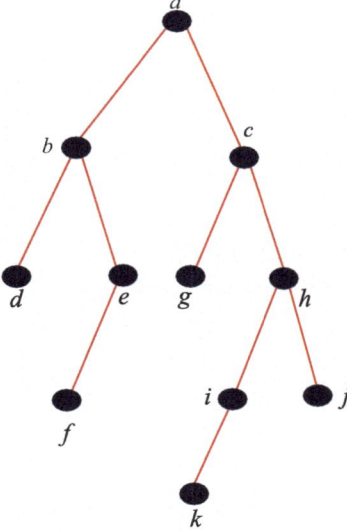

10.14.2.1 Types of Binary Tree

There are five categories of binary tree, each of which is described below.

1. **Full (or strict) binary tree (Fig. 10.27)**

Each node must contain either 0 or 2 children. Put another way, each node except the leaf nodes must contain 2 children.

2. **Complete binary tree (Fig. 10.28)**

The nodes are filled from left to right—every level, except the last one must be full—the nodes of the last level must be as left at possible.

3. **Balanced binary tree (Fig. 10.29)**

The left and right trees differ by at most by one node.

4. **Perfect binary tree (Fig. 10.30)**

All the internal nodes have 2 children, and all the leaf nodes are at the same level. Note that a perfect binary tree is a special instance of a balanced binary tree.

5. **Degenerate binary tree (Figs. 10.31 and 10.32)**

Fig. 10.27 A full binary tree

Fig. 10.28 A complete binary tree

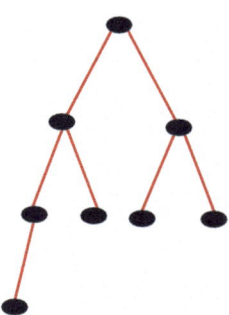

10.14 Trees

Fig. 10.29 Balanced binary trees

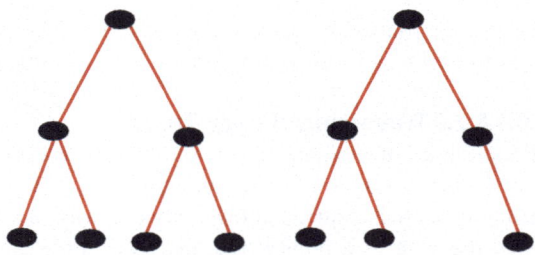

Fig. 10.30 A perfect binary tree

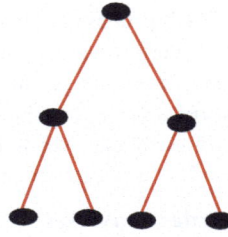

Fig. 10.31 A left-skewed degenerate binary tree

Fig. 10.32 A right-skewed degenerate binary tree

All the internal nodes have only one child. The graphs in Figs. 10.32 and 10.32 are referred to left-skewed and right-skewed binary trees respectively.

10.14.2.2 Traversing Binary Trees

If each node of a binary tree is thought of as representing a piece of data, then a binary tree is one way of organizing data: it is a **data structure**. A simple data structure such as a queue allows only one way of traversing it, namely sequentially.

In the case of a binary tree, there are three algorithms we can use to traverse it—each has its advantages and disadvantages.

Each process entails visiting the root and all nodes in the left hand tree and the right hand tree—as we shall see, the difference between each lies in the order in which this is done. In each case, the process is *recursive*. As each tree is visited, the rules are applied to that tree. Examples should make this clear.

In each case we will use as our example the binary tree shown in Fig. 10.33, which is a perfect binary tree with 15 nodes.

Inorder Traversal

1. Traverse the left subtree.
2. Visit the root.
3. Traverse the right subtree.

Let's apply this to the tree in Fig. 10.33. We start with the left subtree—that is the subtree whose root is *b*. We apply the inorder rules to this tree so we traverse the left subtree of this tree, which is the subtree whose root is *c*. The left subtree of this tree has the root *d*, but this is a leaf-node, so there is no left sub-tree of this tree, and we therefore visit the root *d*.

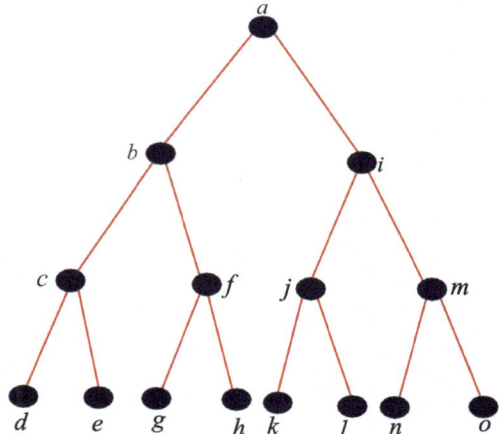

Fig. 10.33 A perfect binary tree with 15 nodes

10.14 Trees

Having visited the root of this subtree, according to the inorder rules, we visit the root, c. We now traverse the right-hand subtree; again this ends with a leaf node, so we visit the node e.

We have now traversed the entire subtree whose root is c—that is to say the left subtree of the tree whose root is b. So now we visit the root of this tree, b. We now visit the right subtree of this tree applying the same rules so we visit g, f and h in that order.

Now we can visit the root of the whole tree, a.

So far we have visited the nodes in this order: d, c, e, b, g, f, h, a

We now apply exactly the same procedure to the right subtree (the one whose root is i). This will mean visiting the nodes in the order k, j, l, i, n, m, o.

Putting these together we get a traversal path of: $d, c, e, b, g, f, h, a, k, j, l, i, n, m, o$

Preorder Traversal

1. Visit the root.
2. Traverse the left subtree.
3. Traverse the right subtree.

Again, we'll apply this to the tree in Fig. 10.33. We visit the root, a. We then traverse the left subtree, namely the one whose root it b. We visit the root, b. We then traverse left subtree of this tree, the one whose root is c. We visit the root, c. We visit the left subtree of this, whose root is d. We visit d. There is no left subtree of this tree, so we visit the right subtree, and visit the root e. We have now traversed the entire left subtree of the tree with root b, so we traverse its right subtree with the same rules, so we visit f, g, h in that order.

So now we must traverse the right subtree, and, following the same rules, we visit i, j, k, l, followed by m, n, o.

So our traversal path is: $a, b, c, d, e, f, g, h, i, j, k, l, m, n, o$

Postorder Traversal

1. Traverse the left subtree.
2. Traverse the right subtree.
3. Visit the root.

Referring again to Fig. 10.33, we start by traversing the left subtree, the one with root b, We traverse the left subtree of this, following the same rules, so we traverse the subtree starting with c. The left subtree of this is the one with root d, which has no left tree and no right tree. So we visit the root, d. We must now traverse the right subtree of the tree with root c. Its left subtree is the one with root e, which again has no subtrees, so we visit e. We then visit the root, c.

We must now visit the right subtree of b, so following the same rules, we visit g, h, f. We then visit the root, b.

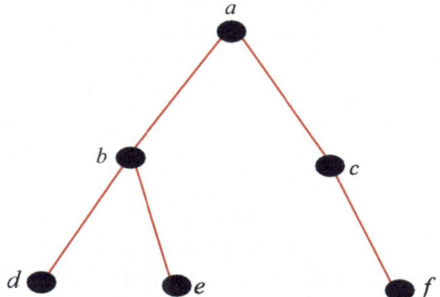

Fig. 10.34 Worked example 6

That completes the traversal of the left subtree, and we now traverse the right subtree in the same way: we visit $k, l,$ j followed by n, o, m, i.

Finally we visit the root of the tree, a.

This gives the following path: $d, e, c, g, h, f, b, k, l, j, n, o, m, i, a$

Worked Example 6
Consider the binary tree shown in Fig. 10.34. Provide the path obtained by using:

(a) Inorder traversal
(b) Preorder traversal
(c) Postorder traversal

Solution

(a) Inorder: d, b, e, a, c, f
(b) Preorder: a, b, d, e, c, f
(c) Postorder: d, e, b, f, c, a.

10.15 Planar Graphs

A **planar** graph is a graph that can be drawn in a plane without any of the edges crossing. Consider the graph shown in Fig. 10.35a. This is a planar graph because it can be redrawn as in Fig. 10.35b. Each particular representation of a planar graph is called a **map**.

A map divides the graph into a number of bounded areas known as **regions**. Figure 10.36 shows a planar graph divided into regions.

10.15 Planar Graphs

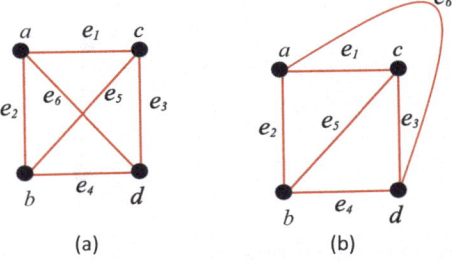

Fig. 10.35 Two maps representing the same planar graph

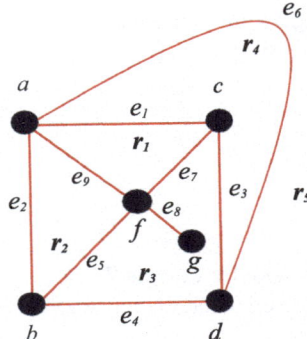

Fig. 10.36 A planar graph with five regions

You will notice that the area outside the graph, r_5 in this case, counts as one region. This is known as an **infinite region**.

The degree of a region refers to the sum of the edges that border the region. In the case of an infinite region, we count the number of edges that bound the graph. Thus

$$deg(r_1) = 3 \quad deg(r_2) = 3 \quad deg(r_3) = 4 \quad deg(r_4) = 3 \quad deg(r_5) = 3$$

The following theorem holds:
The sum of the degrees of a region of a map equals twice the number of edges.
In the above case, the sum of the degrees is 16, and the number of edges is 8.

Euler's Formula

If V is the number of vertices in a map, E is the number of edges, and R is the number of regions, then:

$$V - E + R = 2$$

In the above example:

$$V = 6 \quad E = 9 \quad R = 5$$
$$V - E + R = 6 - 9 + 5 = 2$$

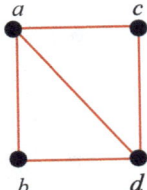

Fig. 10.37 Worked example 7

Fig. 10.38 Solution to worked example 7

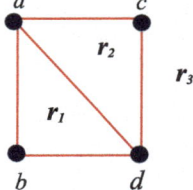

Worked Example 7
Consider the planar graph shown in Fig. 10.37:

(a) Identify the regions.
(b) Give the degree of each region.
(c) Show that Euler's formula holds for this graph.

Solution
See Fig. 10.38.

(a) $deg(r_1) = 3 \quad deg(r_2) = 3 \quad deg(r_3) = 4$
(b) $V = 4 \quad E = 5 \quad R = 3$

$$V - E + R = 4 - 5 + 3 = 2.$$

10.16 Directed Graphs

A **directed graph**—also known as a **digraph**—is a graph, or multigraph, in which the edges have a direction. In the case of a directed graph each edge is an *ordered pair* of vertices.

You briefly encountered digraphs in Chap. 3, where you saw that a digraph represents a particular relation on a set. Another example is presented in Fig. 10.39.

Fig. 10.39 A directed graph

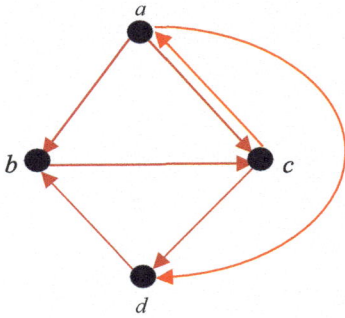

Formally, we can say that a digraph, *G*, consists of a set of vertices, *V*, and a set of ordered pairs, *E*:

$$G = (V, E)$$

In Fig. 10.28:

$$V = \{a, b, c, d\} \quad E = \{(a, b), (a, c), (a, d), (b, b), (b, c), (b, d), (c, d)\}$$

In the case of a directed graph, rather than talk about the degree of a vertex, we talk about the **indegree** and **outdegree**. For example, in Fig. 10.28:

$$indeg(a) = 1 \quad indeg(b) = 2 \quad indeg(c) = 2 \quad indeg(d) = 2$$
$$outdeg(a) = 3 \: outdeg(b) = 1 \: outdeg(c) = 2 \: outdeg(d) = 1$$

It holds that:

The sum of the indegrees of a vertex equals the sum of the outdegrees, which both equal the number of edges.

In Fig. 10.40 we see an adjacency matrix for the directed graph in Fig. 10.39. In the case of a digraph, each entry in the matrix indicates whether there is a path of length 1 from one particular vertex to another. In the example below the direction is from the row to the column.[2] A zero indicates that there is no connection. Some annotations have been added to make it clear.

In the case of a directed graph it can be shown that, for an adjacency matrix *A*, A^2 will show the number of paths of length 2 from a particular vertex to another, A^3 will show the number of paths of length 3 and so on.

In general, for a directed graph with the adjacency matrix *A*, the number of paths of length *k* for each entry of the matrix is given by A^k.

---

[2] In some texts you will find that the direction is from column to row.

Fig. 10.40 Adjacency matrix for a directed graph

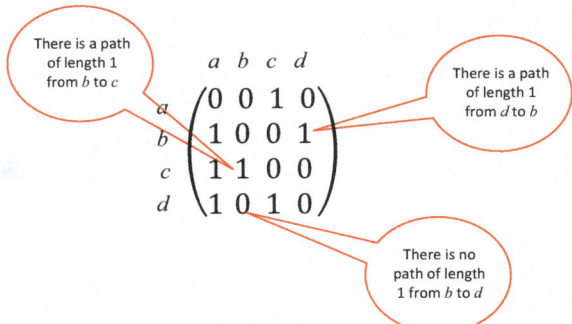

In the above case, referring to the matrix as *A*, we find (using the rules you learnt for matrix multiplication in Chap. 7 (or using an online calculator) that:

$$A^2 = \begin{pmatrix} 1 & 1 & 0 & 0 \\ 1 & 0 & 2 & 0 \\ 1 & 0 & 1 & 1 \\ 1 & 1 & 1 & 0 \end{pmatrix} \begin{matrix} a \\ b \\ c \\ d \end{matrix} \quad A^3 = \begin{pmatrix} 1 & 0 & 1 & 1 \\ 2 & 2 & 1 & 0 \\ 2 & 1 & 2 & 0 \\ 2 & 1 & 1 & 1 \end{pmatrix} \begin{matrix} a \\ b \\ c \\ d \end{matrix}$$

You can verify yourselves that this is indeed correct. For example:

There are two paths of length 2 from *c* to *b*: *cd*, *db* and *ca*, *ab*.

There is one path of length 3 from *d* to *a*: *db*, *bc*, *ca*

Worked Example 8
Consider the digraph shown in Fig. 10.41:

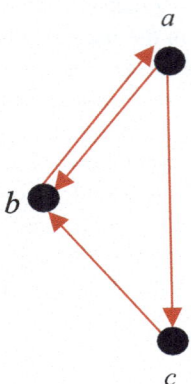

Fig. 10.41 Worked example 8

(a) State the indegree and outdegree for each vertex.
(b) Provide an adjacency matrix, A, showing the number of paths of length 1 from one particular vertex to another.
(c) Derive an adjacency matrix that shows the number of paths of length 2 from one particular vertex to another.

Solution

(a) $indeg(a) = 1 \quad indeg(b) = 2 \quad indeg(c) = 1$
$outdeg(a) = 2 \; outdeg(b) = 1 \; outdeg(c) = 1$

(b) $$A = \begin{pmatrix} 0 & 1 & 0 \\ 1 & 0 & 1 \\ 1 & 0 & 0 \end{pmatrix} \begin{matrix} a \\ b \\ c \end{matrix} \quad \begin{matrix} a\;b\;c \end{matrix}$$

(c) $$A^2 = \begin{pmatrix} 0 & 1 & 0 \\ 1 & 0 & 1 \\ 1 & 0 & 0 \end{pmatrix} \begin{matrix} a \\ b \\ c \end{matrix} \quad \begin{matrix} a\;b\;c \end{matrix}$$

10.17 Application to Computing

Graph theory is extremely important in computer science and related disciplines, because graphs represent networks. They are therefore vital to the field of neural networks and artificial intelligence.

Another example is map applications, which rely on graph theory because navigation problems are modelled as graph problems.

The importance of graph theory in data organization and database development also cannot be over emphasized. For example a binary tree can be used to model a data structure with each node being an item of data. A **binary search tree** is a very important data structure. In this structure data is arranged so that each left subtree contains only nodes with values less than the root and each right subtree contains only nodes with values greater than the root. An example is shown in Fig. 10.42.

A binary search tree enables very efficient search and sort algorithms to be developed, because whole subtrees can be easily discarded without having to be traversed.

Other applications include communication over computer networks, computational flow, and the development of computer hardware.

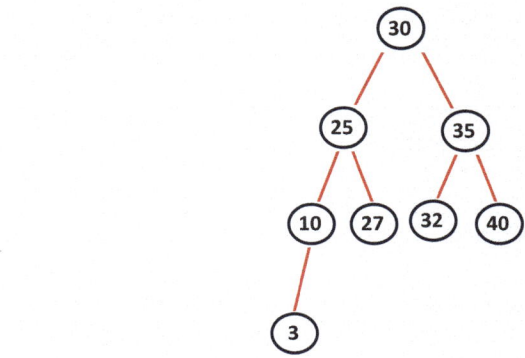

Fig. 10.42 A binary search tree

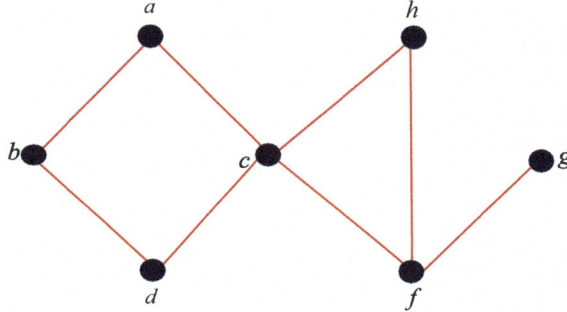

Fig. 10.43 Exercise 1

10.18 Exercises

1. Consider the graph shown in Fig. 10.43:
   (a) State the value of:

   $$deg(a)\ deg(b)\ deg(c)\ deg(d)\ deg(f)\ deg(g)\ deg(h)$$

   (b) Show that the *sum of degrees of vertices* theorem holds for the above graph.
   (c) State the value of:

   $$d(b,g)\ d(a,d)\ d(h,b)$$

   (d) State the value of:

   $$e(a)\ e(b)\ e(c)\ e(d)\ e(f)\ e(g)\ e(h)$$

   (e) Referring to the above graph as G, give the values of:

   $$radius(G)\ diameter(G)$$

Fig. 10.44 Exercise 2

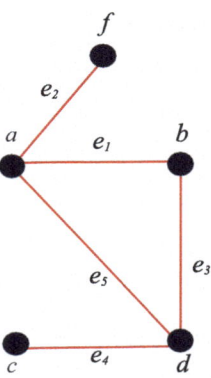

2. Consider the graph shown in Fig. 10.44:
   Now consider the following path: $a, e_1, b, e_3, d, e_4, c$
   State whether this path is:
   (a) a simple path;
   (b) a trail;
   Now consider the following path: $a, e_1, b, e_3, d, e_5, a$.
   State whether this path is:
   (c) a circuit;
   (d) a cycle.
3. Consider the three graphs shown in in Fig. 10.45, and identify if any pairs of them are isomorphic:
4. State whether the graph shown in 10.46 is:
   (a) Traversable;
   (b) Eulerian;
   (c) Hamiltonian.
5. Provide an adjacency matrix for the weighted graph shown in Fig. 10.47.
6. Use Kriskal's algorithm to find the minimum spanning tree for the weighted graph shown in Fig. 10.48.
7. Consider the binary tree shown in Fig. 10.49. Provide the path obtained by using:
   (a) Inorder traversal

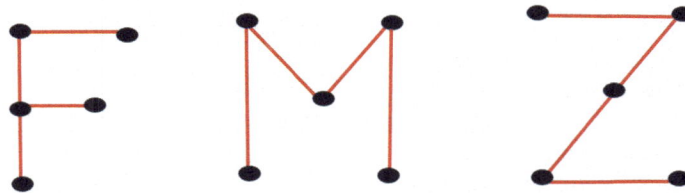

Fig. 10.45 Exercise 3

Fig. 10.46 Exercise 4

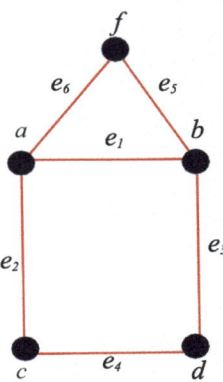

Fig. 10.47 Exercise 5

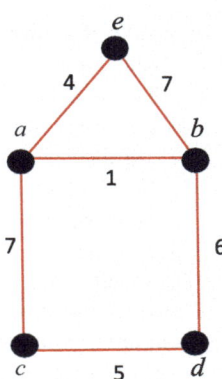

Fig. 10.48 Exercise 6

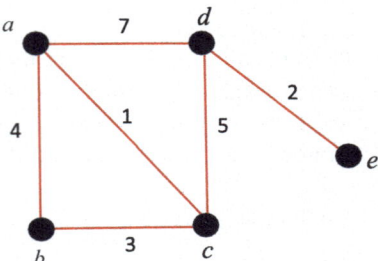

    (b) Preorder traversal
    (c) Postorder traversal
8. Consider the planar graph shown in Fig. 10.50:
    (a) Identify the regions.
    (b) Give the degree of each region.
    (c) Show that Euler's formula holds for this graph.
9. Consider the digraph shown below in Fig. 10.51:

Fig. 10.49 Exercise 7

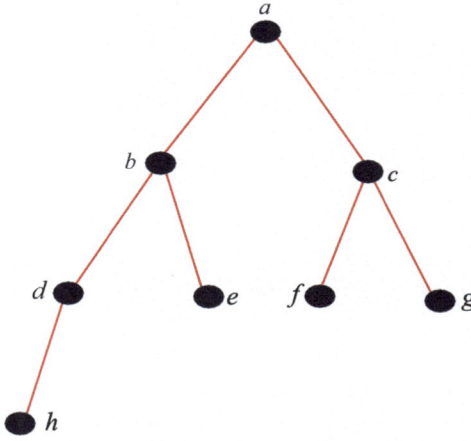

Fig. 10.50 Exercise 8

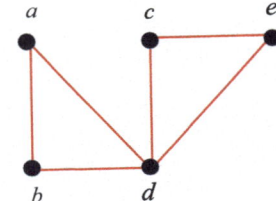

Fig. 10.51 Exercise 9

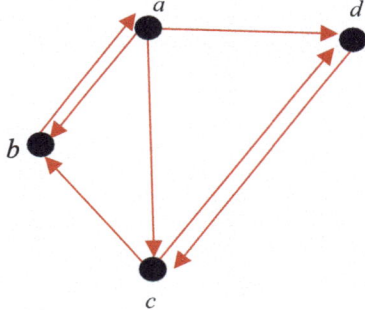

(a) State the indegree and outdegree for each vertex.
(b) Provide an adjacency matrix, A, showing the number of paths of length 1 from one particular vertex to another.
(c) Derive an adjacency matrix that shows the number of paths of length 2 from one particular vertex to another.

Solutions to Exercises

11.1 Chapter 1

1. M is the set of natural numbers greater than or equal to 50.
2. $A = \{x \in \mathbb{Z} | - 5 < x < 5\}$
3. (a) Finite   (b) Infinite   (c) Finite
4. (a) True   (b) False   (c) True   (d) True   (e) False   (f) True
5. (a)
       (i) $A \cap B = \{\text{APPLE, ORANGE}\}$
       (ii) $B \cup C = \{\text{APPLE, MANGO, ORANGE, GRAPE, CHERRY}\}$
       (iii) $A \backslash B = \{\text{PEAR, BANANA, PLUM, LEMON}\}$
       (iv) $B \cap D = \emptyset$
       (v) $B \times D = \{(\text{APPLE, BANANA}), (\text{MANGO, BANANA}), (\text{ORANGE, BANANA})\}$
       (vi) $n(C) = 3$
   (b) $\overline{A} = \{\text{MANGO, GRAPE, CHERRY, PINEAPPLE}\}$

6.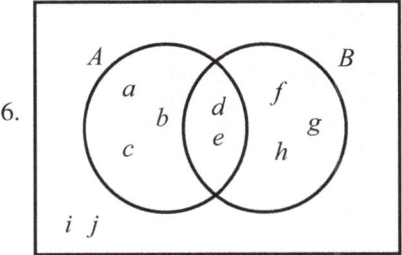

7. $A \Delta B = A \backslash B \cup B \backslash A$
       = {pear, banana, plum, lemon} ∪ {mango}
       = {pear, banana, plum, lemon, mango}

8. (a)
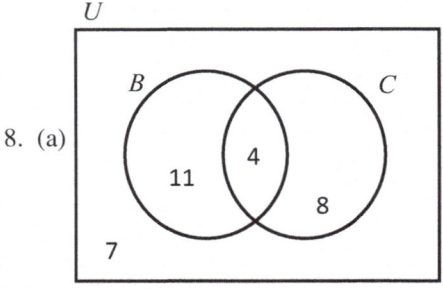

(b)
   (i) $n(B \cap C) = 4$
   (ii) $n(B \cup C) = 23$
   (iii) $n(B \backslash C) = 11$
   (iv) $n(\overline{B \cup C}) = 7$

9.
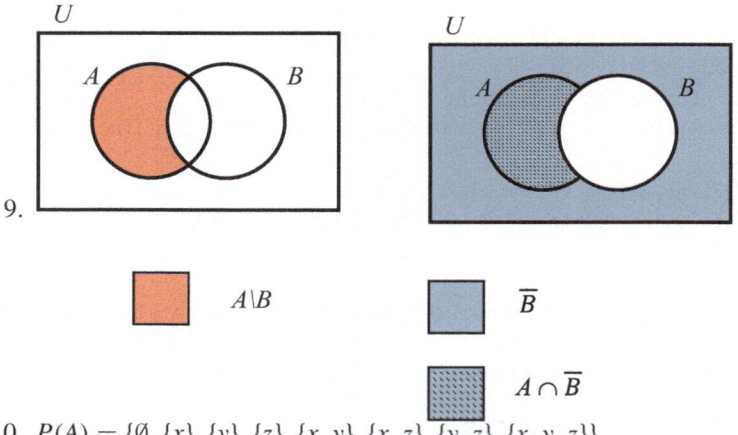

10. $P(A) = \{\emptyset, \{x\}, \{y\}, \{z\}, \{x, y\}, \{x, z\}, \{y, z\}, \{x, y, z\}\}$
11. (a) Number of elements in the power set $= 2^4 = 16$
    (b) Number of proper subsets contained in the power set $= 15$

11.2 Chapter 2

1. $\begin{aligned} A \cup (\overline{A} \cap B) &= (A \cup \overline{A}) \cap (A \cup B) && \text{DistributiveLaw} \\ &= U \cap (A \cup B) && \text{ComplementLaw} \\ &= A \cup B && \text{IdentityLaw} \end{aligned}$

2. $$\begin{aligned} B \cap \overline{(A \cap B)} &= B \cap (\overline{A} \cup \overline{B}) & \text{De Morgan's Law} \\ &= (B \cap \overline{A}) \cup (B \cap \overline{B}) & \text{Distributive Law} \\ &= (B \cap \overline{A}) \cup \emptyset & \text{Complement Law} \\ &= B \cap \overline{A} & \text{Identity Law} \end{aligned}$$

3. (a) $$\begin{aligned} B \cup (A \backslash B) &= B \cup (A \cap \overline{B}) & \text{From previous chapter} \\ &= (B \cup A) \cap (B \cup \overline{B}) & \text{Distributive Law} \\ &= (B \cup A) \cap U & \text{Complement Law} \\ &= B \cup A & \text{Identity Law} \end{aligned}$$

(b)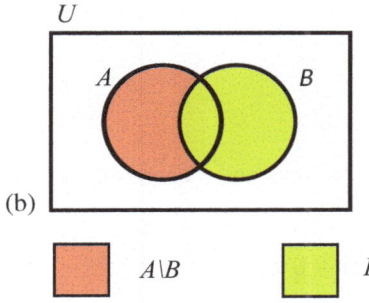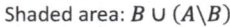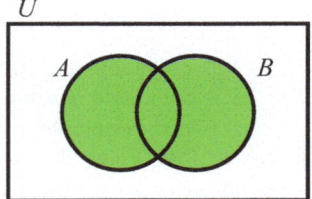

Shaded area: $B \cup (A \backslash B)$ Shaded area: $B \cup A$

4. (a) All of them   (b) −335, 2, 0   (c) 2, 0
5. 7.5 because it can be represented as $\frac{15}{2}$
   $\sqrt{25}$ because it can be represented as $\frac{5}{1}$
6. In each case, remember that $i^2 = -1$.

   (a) $\begin{aligned} i(3+i) &= 3i + i^2 \\ &= -1 + 3i \end{aligned}$

   (b) $\begin{aligned} (2-3i)(4+i) &= 8 + 2i - 12i - 3i^2 \\ &= 11 - 10i \end{aligned}$

   (c) $\begin{aligned} (3-i)^2 &= 9 - 6i + i^2 \\ &= 8 - 6i \end{aligned}$

(d)
$$(1+i)^3 = (1+i)(1+i)^2 = (1+i)(1+2i+i^2)$$
$$= (1+i)(1+2i-1)$$
$$= 2i(1+i)$$
$$= 2i + 2i^2$$
$$= -2 + 2i$$

(e) Multiply numerator and denominator by $(1-i)$

$$\frac{3+4i}{1+i} = \frac{(3+4i)(1-i)}{(1+i)(1-i)}$$
$$= \frac{7+i}{1-i^2} = \frac{7+i}{2}$$

7. A is countable and infinite.

B is countable and finite.

S is countable and finite.

T is non-countable and infinite.

8. (a) We know that subtraction is not an associative operation, so natural numbers under subtraction is not a group.

(b) **Is there an identity element?**

The identity element is 0 because $0 + 0 = 0$.

Are there inverses?

$0 + 0 = 0$, so there is an inverse for each element (there is in fact only one element).

Is the operation associative?

Addition is associative with integers.

Is there closure?

It is closed because the result of $0 + 0$ is in the group.

Therefore the set $\{0\}$ under addition is a group.

(c) **Is there an identity element?**

The identity element is 1 because $1 \times 1 = 1$.

Are there inverses?

$1 \times 1 = 1$, so there is an inverse for each element (there is in fact only one element).

Is the operation associative?

Addition is associative with integers

Is there closure?

It is closed because the result of 1×1 is in the group.

Therefore the set $\{1\}$ under multiplication is a group.

9. **Is there an identity element?**

The identity element is 0.

Are there inverses?

There are no inverses apart from zero.

Is the operation associative?

Addition is associative with integers.

Is there closure?
It is closed because the result of adding any two natural numbers is always a natural number.

Therefore $\mathbb{N}_0$ under addition has an identity, closure and associativity so is a monoid but not a group.

10. **Is there an identity element?**
There is no identity element (0 is not in the group).
Are there inverses?
There are no inverses because there is no identity element.
Is the operation associative?
Addition is associative with integers.
Is there closure?
It is closed because the result of adding any two natural numbers is always a natural number.

Therefore $\mathbb{N}_1$ under addition has associativity and closure and is a semigroup but not a group.

11.3 Chapter 3

1. (a) $R = \{(\text{MARY}, \text{SIDCUP LIONS}), (\text{HADIYA}, \text{SIDCUP LIONS}),$
    $(\text{LINDA}, \text{PENGE TIGERS}), (\text{LINDA}, \text{GRAVESEND BEARS})\}$
   (b) Mary plays for Sidcup Lions

2.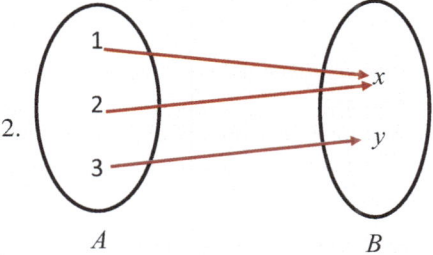

3. $R^{-1} = \{(2, a), (9, d), (4, b), (7, c), (1, a)\}$
4.
    (a) It is not symmetric because if $a < b$, then $b \not< a$. For example 4 is less than 5, but 5 is not less than 4.
    (b) It is not reflexive because in no case is $a < a$.
    (c) It is transitive, because in every case if $a < b$ and $b < c$, then $a < c$. For example 3 is less then 5 and 5 is less than 6 – of course 3 is also less than 6, and this is true in every case.

5.
    (a) It is not reflexive because $1 \not\mathcal{R} 1$.
    (b) It is not symmetric because, for example, $1 \mathcal{R} 3$ but $3 \not\mathcal{R} 1$.
    (c) It is transitive because in every case if $a \mathcal{R} b$ and $b \mathcal{R} c$ then $a \mathcal{R} c$.

11.4 Chapter 4

1.
   (a) This is not a function because d maps on to two elements in the codomain.
   (b) This is a function.
   (c) This is not a function because b has no image in the codomain.
2. (a) $f(u) = 1$ (b) $f(v) = 2$ (c) $f(w) = 2$ (d) $f(x) = 4$
3.
   (a) $f(3) = 4 \times 3^2 - 5 = 4 \times 9 - 5 = 31$
   (b) $f(-1) = 4 \times (-1)^2 - 5 = -1$
   (c) $f(0) = -5$
4.
   (a) $f(2, 0) = 2 \times 2^2 + 3 \times 0 = 8$
   (b) $f(1, -1) = 2 \times 1^2 + 3 \times -1 = -1$
5. $g(3) = 3 \times 3 + 1 = 10$
   $f(10) = 10^3 = 1000$
6. $f : \mathbb{Z} \times \mathbb{Z} \to \mathbb{Z}$
   $f(x, y) = 2(x + y)$
7.
   (a) This is both injective and surjective (it is a bijective function).
   (b) This is surjective, but not injective.
   (c) This is surjective, but not injective.
   (d) This is neither injective nor surjective.

8.
   (a) It is not a surjective function, because the negative numbers in the codomain are not images of any number in the domain.
   (b) It is not an injective function, because more than one element from the domain maps to the same element in the codomain. For example, both 2 and -2 map onto 4.

11.5 Chapter 5

1. P is true
   Q is false
   R is false
   (a) $P \wedge R \equiv T \wedge F \equiv F$
   (b) $P \vee Q \equiv T \vee F \equiv T$
   (c) $Q \vee R \equiv F \vee F \equiv F$
2.
   (a) It is not summer.
   (b) It is summer and Leon is not playing tennis.
   (c) It is not summer or Leon is playing tennis.
   (d) Leon is playing tennis.

11.5 Chapter 5

3.
   (a) $Q \wedge \neg P$
   (b) $P \wedge Q$
   (c) $P \vee \neg Q$
   (d) $\neg(P \vee Q)$

4.

P	Q	$P \wedge Q$	$\neg(P \wedge Q)$	$\neg Q$	$\neg(P \vee Q) \vee \neg Q$
T	T	T	F	F	F
T	F	F	T	T	T
F	T	F	T	F	T
F	F	F	T	T	T

5.

P	Q	$P \vee Q$	$P \Rightarrow (P \vee Q)$
T	T	T	T
T	F	T	T
F	T	T	T
F	F	F	T

6.
   (a) $\neg Q \Rightarrow P$
   (b) $\neg P \Rightarrow \neg\neg Q \equiv \neg P \Rightarrow Q$
   (c) $\neg\neg Q \Rightarrow \neg P \equiv Q \Rightarrow \neg P$

7. $\neg(\neg P \wedge (P \vee Q)) \equiv P \vee \neg(P \vee Q)$
$\equiv P \vee (\neg P \wedge \neg Q)$

8.
$\neg Q \wedge (\neg P \vee Q) \equiv (\neg Q \wedge \neg P) \vee (\neg Q \wedge Q)$
$\equiv (\neg Q \wedge \neg P) \vee F$ Complement Law
$\equiv (\neg Q \wedge \neg P)$ Identity Law

9.
$\neg((P \Rightarrow Q) \wedge Q)) \equiv \neg(P \Rightarrow Q) \vee \neg Q$ De Morgan's Law
$\equiv (P \wedge \neg Q) \vee \neg Q$ Identity 2

10.
$P \Rightarrow (P \vee Q) \equiv \neg P \vee (P \vee Q)$ Identity 1
$\equiv (\neg P \vee P) \vee Q$ $\vee$ is associative
$\equiv T \vee Q$ Complement Law
$\equiv T$ Identity Law

11.

P	Q	$\neg Q$	$P \vee \neg Q$
T	T	F	T
T	F	T	T
T	Undefined	Undefined	T
F	T	F	F
F	F	T	T
F	Undefined	Undefined	Undefined
Undefined	T	F	Undefined
Undefined	F	T	T
Undefined	Undefined	Undefined	Undefined

11.6 Chapter 6

1. (a) (i) Basil is a duck.
    (ii) No animal is a duck.
    (iii) There is at least one animal that is a duck.
    (iv) There exists one and only one animal that is not a duck.
    (v) Not all animals are ducks.
   (b) $\forall x \in A \cdot \neg D(x)$
2. (a) (i) All birds can fly (for every animal, if x is a bird then x can fly).
    (ii) There is at least one bird that can fly (there exists an animal that is a bird and can fly).
    (iii) Jack is a bird or there is at least one animal that can fly.
   (b) (i) $B(\text{Mary}) \equiv \forall x \cdot \neg F(x)$
    (ii) $\forall x \cdot (B(x) \vee \neg F(x))$
    (iii) $\exists! x \cdot (B(x) \wedge \neg F(x))$
3. $\neg \exists x \cdot (\neg P(x) \wedge \neg Q(x)) \equiv \forall x \cdot \neg(\neg P(x) \wedge \neg Q(x))$
4. $\forall x \cdot (P(x) \vee Q(x))$
5. First assign letters to the propositions:
   D: I do the ironing
   C: I have a cup of tea in the afternoon
   T: It is Thursday

We have to show that:

$$\frac{D \Rightarrow C;\ C \Rightarrow T;\ D}{T}$$

11.6 Chapter 6

Proof

    1. $D \Rightarrow C$ Premise
    2. $C \Rightarrow T$ Premise
    3. D Premise
    4. $D \Rightarrow T$ Chain rule on 1 and 2
    5. T Modus Ponens on 3 and 4
    6. Define the following predicates: $C(x)$: x is a cat.
                                                  $T(x)$: x likes watching television.
       the following proposition: P: the moon is made of cheese.
       and the following constants: b: Bernard
                                                     s: Susan

We must prove that: $\dfrac{C(b) \wedge C(s);\ \ T(b);\ \ \exists x \cdot (C(x) \wedge T(x)) \Rightarrow P}{P}$

Proof

    1. $C(b) \wedge C(s)$:   Premise
    2. $T(b)$ Premise
    3. $\exists x \cdot (C(x) \wedge T(x)) \Rightarrow P$ Premise
    4. $C(b)$ AND - Elimination on 1
    5. $C(b) \wedge T(b)$ AND - Introduction on 2 and 4
    6. $\exists x \cdot (C(x) \wedge T(x))$ Existential Generalisation on 5
    7. P Modus Ponens on 3, 6

    7. Define the following predicates: $S(x)$: x is a snake.

                                                       $B(x)$: x can bite.
       and the following proposition: P: Paris is in France.

We must prove that: $\dfrac{S(SAM);\ B(SAM) \Rightarrow P;\ \forall x \cdot (S(x) \Rightarrow B(x))}{P}$

Proof

    1. $S(SAM)$ Premise
    2. $B(SAM) \Rightarrow P$ Premise
    3. $\forall x \bullet (S(x) \Rightarrow B(x))$ Premise
    4. $S(SAM) \Rightarrow B(SAM)$ Universal Instantiation on 3
    5. $B(SAM)$ Modus Ponens on 1, 4
    6. P Modus Ponens on 2, 5

8. **Base step**

Show that it holds when $n = 1$: $2 = 2^2 - 2$

Inductive step

Assume the statement is true for some value $n = k$:

$$2 + 2^2 + 2^3 + 2^4 + \ldots 2^k = 2^{k+1} - 2 \tag{11.1}$$

Now take the sequence up to $k + 1$:

$$2 + 2^2 + 2^3 + 2^4 + \ldots 2^k + 2^{k+1}$$

Substituting from Eq. 11.1 this becomes:

$$\begin{aligned} 2^{k+1} - 2 + 2^{k+1} &= 2 \times 2^{k+1} - 2 \\ &= 2^1 \times 2^{k+1} - 2 \\ &= 2^{(k+1)+1} - 2 \end{aligned}$$

Thus if it true for $n = k$, it is true for $n = k + 1$, and since it is true for $n = 1$, it is true for all $n > 1$

11.7 Chapter 7

1. (a) $A + B = \begin{pmatrix} -2+1 & 3+2 & 5+7 \\ 1+3 & -2+8 & 9+4 \end{pmatrix} = \begin{pmatrix} -1 & 5 & 12 \\ 4 & 6 & 13 \end{pmatrix}$

   (b) $A - B = \begin{pmatrix} -2-1 & 3-2 & 5-7 \\ 1-3 & -2-8 & 9-4 \end{pmatrix} = \begin{pmatrix} -3 & 1 & -2 \\ -2 & -10 & 5 \end{pmatrix}$

   (c) $2A + 3B = \begin{pmatrix} -4 & 6 & 10 \\ 2 & -4 & 18 \end{pmatrix} + \begin{pmatrix} 3 & 6 & 21 \\ 9 & 24 & 12 \end{pmatrix} = \begin{pmatrix} -1 & 12 & 31 \\ 11 & 20 & 30 \end{pmatrix}$

   (d) $A^T = \begin{pmatrix} -2 & 1 \\ 3 & -2 \\ 5 & 9 \end{pmatrix}$

2. $\begin{aligned} (-2 \ 3) &\times \begin{pmatrix} 4 & 1 \\ 2 & 3 \end{pmatrix} \\ &= (-2 \times 4 + 3 \times 2 \quad -2 \times 1 + 3 \times 3) \\ &= (-2 \ 7) \end{aligned}$

3. $A \times B = \begin{pmatrix} 2 \times 1 + 7 \times 0 & 2 \times 5 + 7 \times 2 \\ 1 \times 1 + 3 \times 0 & 1 \times 5 + 3 \times 2 \end{pmatrix} = \begin{pmatrix} 2 & 24 \\ 1 & 11 \end{pmatrix}$

4. $$\begin{aligned}\det(A) &= (1 \times -2)-(4 \times 3)\\ &= -2-12\\ &= -14\end{aligned}$$

5. (a) $\det(A) = 4 \times 3 - 1 \times 2 = 10$

$$A^{-1} = \frac{1}{10}\begin{pmatrix} 3 & -2 \\ -1 & 4 \end{pmatrix} = \begin{pmatrix} 0.3 & -0.2 \\ -0.1 & 0.4 \end{pmatrix}$$

(b) There is no inverse because B is not a square matrix
(c) There is no inverse because $\det(C) = 0$

6. $A \times X = D$
$\therefore X = A^{-1} \times D$

From the previous question we know that: $A^{-1} = \begin{pmatrix} 0.3 & -0.2 \\ -0.1 & 0.4 \end{pmatrix}$

$$\therefore X = \begin{pmatrix} 0.3 & -0.2 \\ -0.1 & 0.4 \end{pmatrix} \times \begin{pmatrix} 2 \\ 2 \end{pmatrix}$$

$$= \begin{pmatrix} 0.3 \times 2 + -0.2 \times 2 \\ -0.1 \times 2 + 0.4 \times 2 \end{pmatrix}$$

$$= \begin{pmatrix} 0.2 \\ 0.6 \end{pmatrix}$$

7.

$$\underset{A}{\begin{pmatrix} 2 & 1 & 4 \\ 0 & 5 & 1 \\ -1 & 4 & 2 \end{pmatrix}} \underset{X}{\begin{pmatrix} x \\ y \\ z \end{pmatrix}} = \underset{B}{\begin{pmatrix} 7 \\ 3 \\ -2 \end{pmatrix}}$$

Using an online calculator:

$$A^{-1} = \begin{pmatrix} \frac{6}{31} & \frac{14}{31} & \frac{-19}{31} \\ \frac{-1}{31} & \frac{8}{31} & \frac{-2}{31} \\ \frac{5}{31} & \frac{-9}{31} & \frac{10}{31} \end{pmatrix} \quad X = A^{-1}B = \begin{pmatrix} \frac{122}{31} \\ \frac{21}{31} \\ \frac{-12}{31} \end{pmatrix}$$

$$x = \frac{122}{31} \quad y = \frac{21}{31} \quad z = \frac{-12}{31}$$

8. Step 1. Write the augmented matrix:

$$\begin{pmatrix} 2 & 5 & | & 21 \\ 1 & 2 & | & 8 \end{pmatrix}$$

Step 2. Perform row operations:

$$\frac{1}{2}R_1 \to R_1 \quad \begin{pmatrix} 1 & \frac{5}{2} & | & \frac{21}{2} \\ 1 & 2 & | & 1 \end{pmatrix}$$

$$R_2 - R_1 \to R_2 \quad \begin{pmatrix} 1 & \frac{5}{2} & | & \frac{21}{2} \\ 0 & -\frac{1}{2} & | & -\frac{5}{2} \end{pmatrix}$$

$$-2R_2 \to R_2 \quad \begin{pmatrix} 1 & \frac{5}{2} & | & \frac{21}{2} \\ 0 & 1 & | & 5 \end{pmatrix}$$

$$R_1 - \frac{5}{2}R_2 \to R_1 \quad \begin{pmatrix} 1 & 0 & | & -2 \\ 0 & 1 & | & 5 \end{pmatrix}$$

The solution is $\boxed{x = -2 \quad y = 5}$

11.8 Chapter 8

1. $5! = 5 \times 4 \times 3 \times 2 \times 1 = 120$

2. $$\frac{8! \times 5!}{4! \times 3!} = \frac{8 \times 7 \times 6 \times 5 \times 4 \times 3 \times 2 \times 1 \times 5 \times 4 \times 3 \times 2 \times 1}{4 \times 3 \times 2 \times 1 \times 3 \times 2 \times 1}$$

$$= 8 \times 7 \times 6 \times 5 \times 5 \times 4$$

$$= 33600$$

11.8 Chapter 8

$$P(n, k) = \frac{n!}{(n-k)!}$$

$$P(10, 3) = \frac{10!}{(10-3)!} = \frac{10!}{7!}$$

3. (a)
$$= \frac{10 \times 9 \times 8 \times 7 \times 6 \times 5 \times 4 \times 3 \times 2 \times 1}{7 \times 6 \times 5 \times 4 \times 3 \times 2 \times 1}$$

$$= 10 \times 9 \times 8 = 720$$

$$C(n, k) = \frac{n!}{(n-k)!k!}$$

$$C(9, 6) = \frac{9!}{(9-6)! \times 6!} = \frac{9!}{3! \times 6!}$$

(b)
$$= \frac{9 \times 8 \times 7 \times 6 \times 5 \times 4 \times 3 \times 2 \times 1}{3 \times 2 \times 1 \times 6 \times 5 \times 4 \times 3 \times 2 \times 1}$$

$$= \frac{9 \times 8 \times 7 \times 6 \times 5 \times 4 \times 3 \times 2 \times 1}{3 \times 2 \times 1 \times 6 \times 5 \times 4 \times 3 \times 2 \times 1}$$

4. This is the same as selecting 4 from 20 where order is significant and there is no repetition, so this is a permutation.
   The correct formula is:

$$P(n, k) = \frac{n!}{(n-k)!} \text{ where } n = 20 \text{ and } k = 4.$$

$$P(20, 4) = \frac{20!}{16!}$$

$$= 20 \times 19 \times 18 \times 17 = 116280$$

5. In both cases, order is not significant and there is no repetition, so this is a combination.
   The correct formula is

$$C(n, k) = \frac{n!}{(n-k)!k!}$$

(a) In this case $n = 10$ and $k = 3$.

$$C(10, 3) = \frac{10!}{7! \times 3!}$$

$$= \frac{10 \times 9 \times 8 \times 7 \times 6 \times 5 \times 4 \times 3 \times 2 \times 1}{7 \times 6 \times 5 \times 4 \times 3 \times 2 \times 1 \times 3 \times 2 \times 1}$$

$$= \frac{10 \times 9 \times 8}{3 \times 2 \times 1} = 120$$

In this case n = 7 and $k = 2$.

$$C(7, 2) = \frac{7!}{5! \times 2!}$$

$$= \frac{7 \times 6 \times 5 \times 4 \times 3 \times 2 \times 1}{5 \times 4 \times 3 \times 2 \times 1 \times 2 \times 1}$$

$$= \frac{7 \times 6}{2 \times 1} = 21$$

6. Order is unimportant, but repetition is allowed.
   The correct formula is: $\frac{(n+k-1)!}{k!(n-1)!}$
   where $n = 10$ and $k = 3$.

$$\frac{(10+3-1)!}{3!(10-1)!} = \frac{12!}{3! \times 9!}$$

$$= \frac{12 \times 11 \times 10 \times 9 \times 8 \times 7 \times 6 \times 5 \times 4 \times 3 \times 2 \times 1}{3 \times 2 \times 1 \times 9 \times 8 \times 7 \times 6 \times 5 \times 4 \times 3 \times 2 \times 1}$$

$$= \frac{12 \times 11 \times 10 \times 9 \times 8 \times 7 \times 6 \times 5 \times 4 \times 3 \times 2 \times 1}{3 \times 2 \times 1 \times 9 \times 8 \times 7 \times 6 \times 5 \times 4 \times 3 \times 2 \times 1}$$

7. Order is important, and repetition is allowed.
   The correct formula is: n^k
   where $n = 6$ and $k = 3$

$$6^3 = 216$$

8. (a) Order is important, and repetition is allowed, so the correct formula to use is n^k, where $n = 6$ and $k = 3$.

$$n^k = 6^3 = 216$$

In this case we must calculate $P(6, 3)$.

$$P(6, 3) = \frac{6!}{3!}$$

$$= \frac{6 \times 5 \times 4 \times 3 \times 2 \times 1}{3 \times 2 \times 1}$$

$$= 120$$

(c) This is the same as putting the 3 at then end and arranging the other 5 digits in the first two slots:

$$P(5,2) = \frac{5!}{3!}$$

$$= \frac{5 \times 4 \times 3 \times 2 \times 1}{3 \times 2 \times 1}$$

$$= 20$$

(d) The number can end in 1 or 4.

From part (c) we see that there are 20 possibilities when one number is fixed at the end. So there are 20 numbers ending with 1 and 20 ending in 4.

So the total number is 40.

9. If there are no restrictions, then the number of ways of arranging the four friends is:

$$4! = 4 \times 3 \times 2 \times 1 = 24$$

If Tracey were to sit at one end, then we just have to arrange the other three. The number of ways of doing this is:

$$3! = 3 \times 2 \times 1 = 6$$

If she sits at the other end there are also 6 ways of arranging the others.

So there are 12 ways of arranging people if Tracey sits at the one or other end.

So the number of ways of arranging them if Tracey does *not* sit at the end is:

$$24 - 12 = 12$$

10.

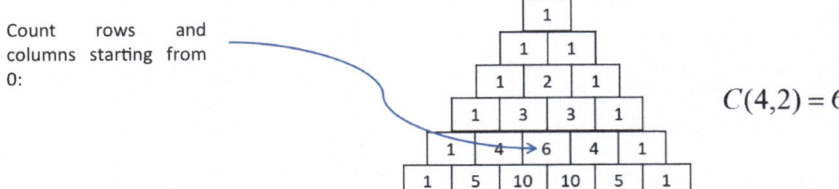

$$C(4,2) = \frac{4!}{2! \times 2!}$$

$$= \frac{4 \times 3 \times 2 \times 1}{2 \times 1 \times 2 \times 1} = 6$$

11. $(a+b)^n = {}^nC_0 a^n b^0 + {}^nC_1 a^{n-1} b^1 + {}^nC_2 a^{n-2} b^2 + \ldots + {}^nC_{n-1} a^1 b^{n-1} + {}^nC_n a^0 b^n$
    In this case: $n = 4$, $a = 2x$ and $b = -y$

$$(2x - y)^4 = {}^4C_0 (2x)^4 (-y)^0 + {}^4C_1 (2x)^3 (-y)^1 + {}^4C_2 (2x)^2 (-y)^2 + {}^4C_3 (2x)^1 (-y)^3 + {}^4C_4 (2x)^0 (-y)^4$$
$$= 16x^4 + 4(8x^3)(-y) + 6(4x^2)(y^2) + 4(2x)(-y^3) + (y^4)$$
$$= 16x^4 - 32x^3 y + 24x^2 y^2 - 8xy^3 + y^4$$

12. $(a+b)^n = \sum_{k=0}^{n} nC_k a^{n-k} b^k$
    In this case, $a = x$ and $b = 2y$
    $n = 6$, and because we start at zero, the third term will be given when $k = 2$.

The third term is therefore:
$${}^6C_2 (x^4)(2y)^2$$
$$= 15 (x^4)(4y^2)$$
$$= 60 x^4 y^2$$

11.9 Chapter 9

1. (a) $S = \{\text{YELLOW, BLUE, GREEN, RED, BLACK}\}$
   (b) $E = \{\text{BLUE, BLACK}\}$
   (c) $n(S) = 5 \quad n(E) = 2$
   (d) $P(E) = \frac{n(E)}{n(S)} = \frac{2}{5}$
2. (a) $S = \{1, 2, 3, 4, 5, 6, 7, 8, 9, 10, 11, 12\}$
   (b) $A = \{1, 2, 3, 4, 5, 6, 7\} \quad B = \{1, 3, 5, 7, 9, 11\} \quad C = \{4, 8, 12\}$
   (c) $A \cap B = \{1, 3, 5, 7\} \quad A \cap C = \{4\} \quad B \cap C = AE$
   (d) B and C are mutually exclusive
       $P(B \cup C) = P(B) + P(C) - P(B \cap C)$
   (e)
   $$= \frac{n(B)}{n(S)} + \frac{n(C)}{n(S)} - \frac{n(B \cap C)}{n(S)}$$
   $$= \frac{6}{12} + \frac{3}{12} - \frac{0}{12} = \frac{3}{4}$$

(f)
$$P(A \cup C) = P(A) + P(C) - P(A \cap C)$$
$$= \frac{n(A)}{n(S)} + \frac{n(C)}{n(S)} - \frac{n(A \cap C)}{n(S)}$$
$$= \frac{7}{12} + \frac{3}{12} - \frac{1}{12} = \frac{9}{12} = \frac{3}{4}$$

3. $p(A) = 0.6$
$p(B) = 0.4$
$p(A \cap B) = 0.2$

Apply the addition rule:

$$p(A \cup B) = p(A) + p(B) - p(A \cap B)$$
$$p(A \cup B) = 0.6 + 0.4 - 0.2 = 0.8$$

4. Let S be the sample space: $n(S) = 15$

Let A be the event of drawing an ace: $n(A) = 3$
Let B be the event of drawing a black card: $n(B) = 8$
There are 2 cards that are black and are aces: $n(A \cap B) = 2$

$$P(A \cup B) = P(A) + P(B) - P(A \cap B)$$
$$= \frac{n(A)}{n(S)} + \frac{n(B)}{n(S)} - \frac{n(A \cap B)}{n(S)}$$
$$= \frac{3}{15} + \frac{8}{15} - \frac{2}{15} = \frac{3}{5}$$

5. (a) Let the probability of the spinner landing on orange be x. The probability of it landing on green is $2x$. The probability of it landing on blue is $3x$. The probability of it landing on yellow is $4x$. The probability of it landing on red is $10x$.

The probability distribution is:

$$P(\text{RED}) = 10x \quad P(\text{YELLOW}) = 4x \quad P(\text{BLUE}) = 3x$$
$$P(\text{GREEN}) = 2x \quad P(\text{ORANGE}) = x$$

The events are mutually exclusive, so the probabilities must up to 1:

$$10x + 4x + 3x + 2x + x = 1$$
$$20x = 1$$
$$x = 0.05$$

The probability distribution is:

$$P(\text{RED}) = 0.5 \quad P(\text{YELLOW}) = 0.2 \quad P(\text{BLUE}) = 0.15$$
$$P(\text{GREEN}) = 0.1 \quad P(\text{ORANGE}) = 0.05$$

(b) The probability of the spinner landing on blue or green $= P(\text{BLUE}) + P(\text{GREEN})$
   $= 0.25$

6. The table shows that there are 9 possible outcomes:

Outcome	Spinner 1	Spinner 2
1	RED	RED
2	RED	BLUE
3	RED	YELLOW
4	BLUE	RED
5	BLUE	BLUE
6	BLUE	YELLOW
7	YELLOW	RED
8	YELLOW	BLUE
9	YELLOW	YELLOW

If S is the sample space, then $n(S) = 9$

There are 5 outcomes (outcomes 3, 6, 7, 8 and 9) in which at least one spinner lands on yellow.

If E is the event of at least one spinner landing on yellow, then $n(E) = 5$
Therefore: $P(E) = \frac{n(E)}{n(S)} = \frac{5}{9}$

7. The possibilities are:

Spinner 1	Spinner 2	Number of yellows
RED	RED	0
RED	YELLOW	1
RED	BLUE	0
YELLOW	RED	1
YELLOW	YELLOW	2
YELLOW	BLUE	1
BLUE	RED	0

(continued)

(continued)

Spinner 1	Spinner 2	Number of yellows
BLUE	YELLOW	1
BLUE	BLUE	0

Probability distribution:

No of yellows thrown, x	0	1	2
Number of occurrences	4	4	1
$P(X = x)$	$4/9$	$4/9$	$1/9$

Therefore: $E(x) = 0 \times \frac{4}{9} + 1 \times \frac{4}{9} + 2 \times \frac{1}{9} = \frac{2}{3}$

8. (a) In this case the events are independent.

The probability of picking a blue ball is: $\frac{4}{15}$

The probability of picking a red ball is: $\frac{6}{15}$
The probability of picking a blue ball followed by a red ball is:

$$\frac{4}{15} \times \frac{6}{15} = \frac{4}{15} \times \frac{2}{5} = \frac{8}{75}$$

(b) In this case the events are not independent.

The probability of picking a blue ball is: $\frac{6}{15}$
The probability of picking a red ball is: $\frac{5}{14}$
The probability of picking a blue ball followed by a red ball is:

$$\frac{6}{15} \times \frac{5}{14} = \frac{2}{5} \times \frac{5}{14} = \frac{1}{7}$$

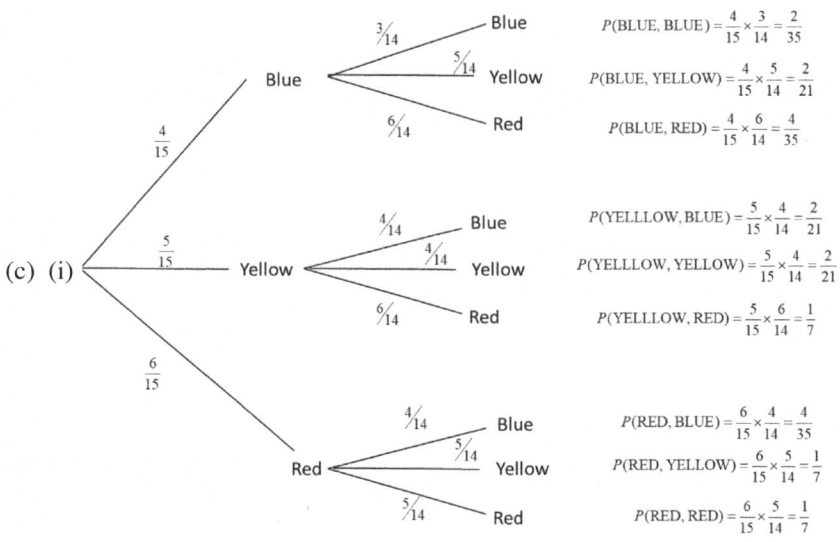

(ii) The probability of picking a red ball followed by a yellow ball OR a yellow ball followed by a blue ball is:

$$\frac{1}{7} + \frac{2}{21} = \frac{3+2}{21} = \frac{5}{21}$$

9. (a) $S = \{(\text{RED, RED}), (\text{RED, WHITE}), (\text{RED, YELLOW}), (\text{WHITE, RED}), (\text{WHITE, WHITE}), (\text{WHITE, YELLOW})\}$
   (b) $E = \{(\text{RED, WHITE}), (\text{WHITE, RED})\}$
   (c) $F = \{(\text{RED, YELLOW})\}$
   (d) $n(S) = 6 \quad n(E) = 2 \; n(F) = 1$
   e) $P(E) = \frac{n(E)}{n(S)} = \frac{1}{3}$
   (f) $P(F) = \frac{n(F)}{n(S)} = \frac{1}{6}$

$P(\text{VOTE}) = 50\% = 0.5$
$P(\text{UNDER 40}) = 30\% = 0.3$
$P(\text{VOTE}|\text{UNDER 40}) = 0.35$

10. $P(\text{UNDER 40}|\text{VOTE}) = \dfrac{P(\text{UNDER 40}) \times P(\text{VOTE}|\text{UNDER 40})}{P(\text{VOTE})}$

$$= \frac{0.3 \times 0.35}{0.5}$$

$$= 0.21$$

11. We use the formula: $P(k \text{ successes in } n \text{ trials}) = {}^nC_k p^k q^{n-k}$

11.9 Chapter 9

$$n = 12$$
$$k = 6$$
In this case: $n - k = 6$
$$p = 0.25$$
$$q = 0.75$$

$$\begin{aligned} P(6 \text{ successes in 12 trials}) &= {}^{12}C_6 \times 0.25^6 \times 0.75^6 \\ &= 924 \times 0.000244 \times 0.177979 \\ &\approx 0.04 \end{aligned}$$

12. We use the formula: $P(k \text{ successes in } n \text{ trials}) = {}^nC_k p^k q^{n-k}$
    $$n = 4$$
    $$k = 3$$
    (a) $n - k = 1$
    $$p = 1/6$$
    $$q = 5/6$$

    $$\begin{aligned} P(3 \text{ successes in 4 trials}) &= {}^4C_3 \times \left(\frac{1}{6}\right)^3 \times \left(\frac{5}{6}\right)^1 \\ &= 4 \times \frac{5}{1296} \approx 0.0031 \end{aligned}$$

    (b) The probability distribution is now: $p(1) = 0.1$

    $$p(2) = 0.1$$
    $$p(3) = 0.1$$
    $$p(4) = 0.1$$
    $$p(5) = 0.3$$
    $$p(6) = 0.3$$

    So now $p = 0.3$, $q = 0.7$

    $$\begin{aligned} P(3 \text{ successes in 4 trials}) &= {}^4C_3 \times 0.3^3 \times 0.7^1 \\ &= 4 \times 0.027 \times 0.7 = 0.0756 \end{aligned}$$

11.10 Chapter 10

1. (a) $\deg(a) = 2 \quad \deg(b) = 2 \quad \deg(c) = 4 \quad \deg(d) = 2$
   $\deg(f) = 3 \quad \deg(g) = 1 \quad \deg(h) = 2$
   (b) The sum of the degrees is 16, which is twice the number of edges, 8.
   (c) $d(b, g) = 4 \quad d(a, d) = 2 \quad d(h, b) = 3$
   (d) $e(a) = 3 \quad e(b) = 4 \quad e(c) = 2 \quad e(d) = 3$
   $e(f) = 3 \quad e(g) = 4 \quad e(h) = 3$
   (e) $radius(G) = 2 \quad diameter(G) = 4$

2. (a) It is a simple path because all vertices are distinct.
   (b) It is a trail because all edges are distinct.
   (c) It is a circuit, because all edges are distinct, and it begins and ends on the same vertex.
   (d) It is a cycle because all vertices, apart from the beginning and end vertices, are distinct. $radius(G) = 2 \quad diameter(G) = 4$

3. **M** and **Z** are isomorphic.

4. (a) It is traversable because it has precisely two vertices (a and b) with odd degree.
   (b) It is not Eulerian, because not all vertices have an even degree.
   (c) It is Hamiltonian, because it contains a circuit that visits each vertex precisely once. For example:

   $$a, e_6, f, e_5, b, e_3, d, e_4, c, e_2, a$$

5. $$\begin{array}{c} \\ a \\ b \\ c \\ d \\ e \end{array} \begin{array}{c} a\ b\ c\ d\ e \\ \begin{pmatrix} 0 & 1 & 7 & 0 & 4 \\ 1 & 0 & 0 & 6 & 7 \\ 7 & 0 & 0 & 5 & 0 \\ 0 & 6 & 5 & 0 & 0 \\ 4 & 7 & 0 & 0 & 0 \end{pmatrix} \end{array}$$

6. Arrange the edges in order of weight:

ac	de	bc	ab	cd	ad
1	2	3	4	5	7

Add edges in order of weight:

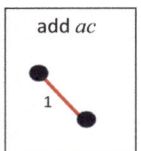 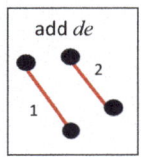 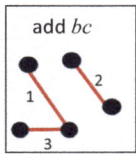

reject *ab* because it would create a cycle

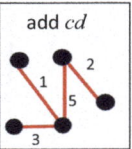

The minimum spanning tree is therefore:

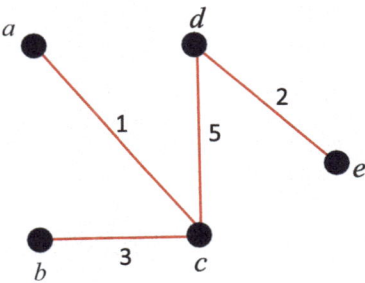

7. (a) Inorder traversal h, d, b, e, a, f, c, g
   (b) Preorder traversal: a, b, d, h, e, c, f, g
   (c) Postorder traversal: h, d, e, b, f, g, c, a

8. (a)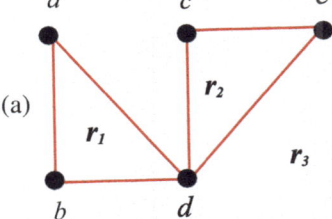

   (b) $deg(r_1) = 3 \quad deg(r_2) = 3 \quad deg(r_3) = 6$
   (c) $V = 5 \quad E = 6 \quad R = 3$

$$V - E + R = 5 - 6 + 3 = 2$$

9. (a) $\quad in\,deg(a) = 1 \quad in\,deg(b) = 2 \quad in\,deg(c) = 2 \quad in\,deg(d) = 2$
$\quad Out\,deg(a) = 3 \quad out\,deg(b) = 1 \quad out\,deg(c) = 2 \quad out\,deg(d) = 1$

(b)
$$A = \begin{matrix} & a\ b\ c\ d & \\ & \begin{pmatrix} 0&1&0&0 \\ 1&0&1&0 \\ 1&0&0&1 \\ 1&0&1&0 \end{pmatrix} & \begin{matrix} a \\ b \\ c \\ d \end{matrix} \end{matrix}$$

Direction is row to column

(c)
$$A^2 = \begin{matrix} & a\ b\ c\ d & \\ & \begin{pmatrix} 1&0&1&0 \\ 1&1&0&1 \\ 1&1&1&0 \\ 1&1&0&1 \end{pmatrix} & \begin{matrix} a \\ b \\ c \\ d \end{matrix} \end{matrix}$$

Index

A
Addition rule, 136
Adjacency matrix, 158
Adjacent vertices, 158
Algebra of propositions, 77–79
Algebra of sets, 22
AND-elimination, 93
AND gate, 81
AND-introduction, 93
AND operator, 66–67
Associative operations, 28

B
Balanced binary tree, 174
Bayes' theorem, 147–150
Bernouilli trial, 150, 151
Bernoulli, 150
Bijective functions, 60–61
Binary relations, 45
Binary trees, 173–178
Binomial expansion, 127
Binomial probability, 150

C
Cardinality, 5
Cartesian product, 10
Chain rule, 92
Circuit, 164
Classes of sets, 16
Combination definition, 120
Combinatorics, 117
Commutative operations, 27
Complement, 9–10
Complete binary tree, 174
Complex numbers, 3, 24–26
Conditional probability, 145–147
Conjunction, 67
Connected graph, 159
Connectives, 66–70
Contradiction, 75–76
Contrapositive, 76–77
Converse, 76–77
Countable sets, 26
Cycle, 164

D
Databases, 46
Degenerate binary tree, 175
Degree of a vertex, 160
De Morgans law, 16, 70–72
Determinant, 105–107
Diameter of a graph, 162
Difference, 9
Digital electronics, 80
Digraph, 180–182
Directed graph, 180–182
Disjunction, 68
Distance between vertices, 160
Domain of discourse, 86

E
Eccentricity, 161
Edge, 158
Empty set, 5
Equivalence classes, 44
Equivalence operator, 73–74
Equivalence relations, 43–44
Euler's number, 23
Eulerian graph, 167
Events, 132
Exclusion principle, 8
EXCLUSIVE OR operator, 76

Existential generalisation, 197
Existential quantifier, 87–88
Expected value, 143–154

F
Factorials, 117–118
Full binary tree, 174
Function composition, 59
Function definition, 49
Function signature, 56

G
Gauss-Jordan elimination method, 113–114
Graph definition, 158
Group definition, 30

H
Hamiltonian circuit, 167
Hamiltonian graph, 167–169
Homeomorphic graph, 165

I
Identity element, 30–31
Identity matrix, 107
Implication operator, 72–73
Independent events, 141
Induction, 96–98
Injective functions, 61
Inorder traversal, 176–177
Integers, 3, 23
Intersection, 8–9
Inverse of a group, 30–31
Inverse of a matrix, 107
Inverse relation, 39
Isomorphic graph, 165

K
Kriskal's algorithm, 171–173

L
Linear equations, 110–111
Logical equivalence, 70, 74
Logical operators, 66–70, 73–74
Logic gates, 80–81

M
Map, 178

Matrix addition, 102
Matrix definition, 101
Matrix inverse, 107–108
Matrix multiplication, 103–105
Matrix operations, 102–105
Matrix subtraction, 102
Mean value, 143
Minimum spanning tree, 171–173
Modus ponens, 91
Modus tollens, 91
Multigraph, 159
Mutual exclusion, 135–136

N
N-ary relations, 45
Natural deduction, 91–96
Natural numbers, 3, 23
Negation, 68
Node, 158
NOT gate, 80
NOT operator, 68

O
One-to-one functions, 60
Onto functions, 60
Order of precedence of logical operators, 74
OR gate, 81
OR-introduction, 94
OR operator, 67–68
Outcome, 132

P
Pascal's Triangle, 126–127
Path, 163–165
Perfect binary tree, 174
Permutation definition, 119
Planar graph, 178–180
Postorder traversal, 177
Power sets, 16–17
Predicate definition, 86
Preorder traversal, 177
Probability definition, 133
Probability distribution, 139–140
Proof by induction, 96–98
Proper subset, 7
Proposition definition, 66

Q
Quantification, 87–90

Index

R
Radius of a graph, 162
Random variables, 142–143
Rational numbers, 3, 23
Real numbers, 3, 24
Reflexive relations, 42–43
Region, 178–180
Relational databases, 46
Relation definition, 37, 40
Row operations, 112
Rule of contraposition, 93
Rule of detachment, 93
Rule of syllogism, 93
Russell's paradox, 17

S
Sample space, 132
Scalar multiplication, 103
Set algebra, 22
Set comprehension, 4
Set definition, 2
Spanning trees, 171–173
Subgraph, 163
Subset, 6
Substitution, 87
Surjective functions, 60
Symmetric difference, 13
Symmetric relations, 42–43

T
Tautology, 75–76
Three-valued logic, 81–82
Trail, 164
Transitive relations, 43
Transposition, 102
Traversable graph, 166–169
Traversing a binary tree, 176–178
Tree diagrams, 145–147
Trees, 170–178
Truth tables, 67–70, 71–74

U
Unconnected graph, 159
Union, 8
Unique existential quantifier, 88
Universal instantiation, 94
Universal quantifier, 87
Universal set, 4–5

V
Venn diagrams, 11–16
Vertex, 158

W
Weighted graph, 169–170

SPRINGER NATURE

GPSR Compliance

The European Union's (EU) General Product Safety Regulation (GPSR) is a set of rules that requires consumer products to be safe and our obligations to ensure this.

If you have any concerns about our products, you can contact us on ProductSafety@springernature.com

In case Publisher is established outside the EU, the EU authorized representative is:

Springer Nature Customer Service Center GmbH
Europaplatz 3
69115 Heidelberg, Germany

The manufacturer's authorised representative in the EU is Springer Nature Customer Service Centre GmbH, Europaplatz 3, 69115 Heidelberg, Germany. If you have any concerns regarding our products, please contact ProductSafety@springernature.com

Printed and bound by CPI Group (UK) Ltd, Croydon, CR0 4YY

26/03/2026

02078953-0004